William Richard Gowers

Clinical lectures on diseases of the nervous system, delivered at the National Hospital for the paralysed and epileptic, London

William Richard Gowers

Clinical lectures on diseases of the nervous system, delivered at the National Hospital for the paralysed and epileptic, London

ISBN/EAN: 9783337229153

Printed in Europe, USA, Canada, Australia, Japan

Cover: Foto ©berggeist007 / pixelio.de

More available books at www.hansebooks.com

ON

DISEASES

OF THE

NERVOUS SYSTEM

DELIVERED AT THE NATIONAL HOSPITAL FOR THE
PARALYSED AND EPILEPTIC, LONDON.

BY

W. R. GOWERS, M.D., F.R.S.,

PHYSICIAN TO THE HOSPITAL; CONSULTING PHYSICIAN TO UNIVERSITY COLLEGE HOSPITAL; FORMERLY PROFESSOR OF CLINICAL MEDICINE IN UNIVERSITY COLLEGE.

PHILADELPHIA:
P. BLAKISTON, SON & CO.,
1012 WALNUT STREET.
1895.

PREFACE.

The following Lectures have been delivered at the National Hospital for the Paralysed and Epileptic. They are reprinted from various English medical journals, with the exception of two lectures, and for permission to reproduce them I am indebted to the J. B. Lippincott Company.

W. R. GOWERS.

Queen Anne St., London,
  June, 1895.

CONTENTS.

	PAGE
I. The Principles of Diagnosis of Diseases of the Nervous System,	9
II. Mistaken Diagnosis,	21
III. Argyria and Syphilis,	34
IV. Syphilitic Hemiplegia,	48
V. Bulbar Paralysis,	71
VI. Facial Paralysis,	89
VII. Facial Contraction after Palsy,	108
VIII. Acute Ascending Myelitis,	119
IX. Locomotor Ataxy, I,	129
X. " " II,	143
XI. The Foot Clonus and its Meaning,	163
XII. Syringo-Myelia,	175
XIII. The Treatment of Muscular Contraction,	188
XIV. The Infantile Causes of Epilepsy, I,	197
XV. " " " " II,	205
XVI. Neuralgia,	214
XVII. Lead Palsy,	232
XVIII. Saturnine Tabes,	242
XIX. Optic Neuritis, I,	252
XX. " " II,	266

CLINICAL LECTURES

ON

DISEASES OF THE NERVOUS SYSTEM

LECTURE I.

SOME GENERAL PRINCIPLES OF THE DIAGNOSIS OF DISEASES OF THE NERVOUS SYSTEM.

Gentlemen:—If we look back over the progress of medical science, three epochs of discovery stand out in special salience, contrasting, in the steepness of the rise in knowledge they present, with the more gradual progress of other branches and of the same subjects at other times. These are, first, the revolution in the conception of diseases of the heart and circulation which Harvey's great discovery entailed; secondly, the penetrating extension of knowledge of all the thoracic diseases which was effected by the invention of the methods of auscultation and percussion, and the discoveries which followed that invention; and, thirdly, the enormous advance in our knowledge of the nervous system and its diseases which the last quarter of a century has witnessed. This has been largely due to the development of microscopical research, in some degree to

Post-graduate Lecture, *Lancet*, Feb. 20, 1892.

the progress of experiment, but very much to the extraordinary increase in the capacity for investigation which the general progress of science has produced, and to the fertile field presented by these diseases for the exercise of that faculty. The results of the application of the chemical and microscopical investigation to the urine, and the transforming revelations made through the invention of the ophthalmoscope and the laryngoscope, are hardly less remarkable, but their range is more limited than those of the progress effected during the three great epochs in discovery that I have mentioned. The last of these concerns us in a special manner, because it has taken place in our own time and is still in active progress. It is to certain consequences of it that I desire to direct your attention to-day—consequences that are of the utmost importance to the practitioner in his daily work. How great, how wide and profound, has been the change that our knowledge of diseases of the nervous system has undergone cannot, indeed, be realized by all members of our profession. A large number of those who studied before the change took place have been unable to follow the new development of knowledge,—the successive stages of which have followed each other with a rapidity almost bewildering, and the difficulty has been enhanced by the freedom with which investigators have revelled in a novel nomenclature, embracing alike the new and the old, in part essential, but in still larger measure superfluous. The exigencies of practice, with its increasing demand on the time and energy of the practitioner, and its unceasing echo of the cry (shall I say of a colleague's daughter?) " Give, give," has made it often impossible for him to attempt, or, attempting to continue, to follow the successive discoveries, Without a knowledge of their scientific features he has generally found it impracticable to use them, and so he has either rested content with that which he learned of old, or has gained a fragmentary knowledge that has served only to mislead him. Those, on the other hand, who have learned the subject as we have it now, cannot realise the change, for they have but one of the terms needed for the comparison; while the large number of practitioners who studied medicine during the period in which the change was taking place, learned a little of the new and a good deal of the old, and acquired a mixture containing many incompatibles, with the common consequence that may be described in terms of metaphor, not inexact, as "mental turbidity."

To clear your minds and get rid, as far as may be, of all that prevents a clear vision of disease, to learn how most usefully to see disease through the clearer media, and, above all, to see it in such a way as shall enable you to treat it with such advantage as is attainable—these, I conceive, are the objects that bring together here the members of this post-graduate course. But I am not sure that the students of to-day, or at least many of them, who have acquired the latest and fullest knowledge of these diseases, have learned, in equal measure, how to use their knowledge, or are free from the need of that training in method which the practitioner feels so keenly. The reason for this, or at least the conditions that make it probable, you will perceive when I tell you the nature of the difficulty that is experienced and the character of the need that is felt.

Let me, however, first give you an illustration of the change that has come over our knowledge of diseases of the nervous system during the last quarter of a century. I remember, twenty-five years ago, listening to a lecture on "chronic spinal meningitis," given by one of the most competent teachers of the day, which presented the latest and clearest opinions on the subject. The malady, as described in that lecture, had no existence. In the case on which it was given, the membranes of the spinal cord were certainly in a perfectly normal state. The symptoms then thought to prove the existence of chronic meningitis are now known to depend solely upon changes within the spinal cord itself. The pathology had been inferred from analogy, but the new means of investigation we have since obtained, coupled with the more extensive and precise observation which I have mentioned as so salient a feature of recent progress, have proved that the analogy was false and the malady quite different from that which was supposed to exist. The symptoms which were then ascribed to chronic meningitis are, indeed, now known to manifest more than one distinguishable disease. But the chief source of practical difficulty is not the mere correction of erroneous inference, and the displacement of the work of fancy by ascertained fact; it depends on the vast increase in our knowledge of the anatomy and physiology of the nervous system and on the application of this to disease; it depends, further, on the revelation, effected by microscopical research, of the extent and variety of degenerative processes in all parts of the nervous system. It depends, lastly, on our ever-widening perception of so-called "functional," really nutritional, diseases for which no anatomical cause has yet been discovered, which have been distinguished or recognised by means of (1) the increased exactness of observation, (2) the aid received through the differentiation of processes formerly confused with these, but now known to depend on structural change, and (3) in consequence of the special indications afforded by the discoveries now being made of the vast range of influence of toxic agents on the nervous system. Can we wonder, then, that our knowledge has been changed, even to the point of transformation, and that those familiar only with the old should be almost lost in the mazes of the new? The practical effect of these changes is to render the customary methods of diagnosis wholly inapplicable to a large proportion of cases of disease of the nervous system, to render it useless, or more than useless in the case of a large number of the nutritional affections that are called "functional," and of the structural maladies that are called "degenerative," as well as many of the maladies that depend on the visible and conspicuous changes that are called "organic."

The student has to acquire his knowledge of all diseases by the method of "types," and the same method is employed in the practical diagnosis of most other maladies. Indeed, in the case of many, no other system of recognising the nature of a case could be employed than that of observing the correspondence of the symptoms, in character and course, with those of a certain form which is known by a special name. But when we turn to the diseases of the nervous system that I have indicated, we meet with a series of maladies the variety of which is almost infinite. When we consider the various structures comprised in the "nervous system," their diversity of function, the extremely numerous combinations of structure and function that may be deranged at the same time, all mutually acting and reacting to produce the manifestation of disorder; when we consider further the variety of nutritional and degenerative processes to which these structures are liable, and also the degree and extent to which they are influenced by toxic blood states—agencies that we are only beginning to perceive in their nature and diversity,—we can readily understand that it not only may be, but that it must be, true that the diseases we meet with are so numerous and so diverse as entirely to baffle an attempt to range them under definite types, that can be designated by definite names. But in the series they form, there are here and there aggregations, as it were, in which the morbid process corresponds, at least in its chief features, varying only in minor points. These we treat as types, and call by names, and by them the student learns the general features of the classes of disease in which they occur; but they seldom stand alone or are marked off by any well-defined boundary. Intermediate cases connect adjacent groups—cases that combine the features of two or several, or present some characteristics of one and some of another, and very often symptoms that cannot be recognised as pertaining to any recognised type. Of these forms, numerous as they are, the student for the most part remains ignorant. He has enough to do to become acquainted with the more familiar groups, and the method by which he learns the features of these cannot take in the intermediate varieties. When he proceeds to practical work he soon finds himself confronted by cases the like of which he has never heard of, and which he cannot find described in any text-book of these diseases. If he attempts to identify them, to bring them under some recognised type, and to give them some definite name, he is at once and completely baffled. He is baffled of necessity, for the cases belong to no type, and no name can be given to them that does not involve more error than truth. It cannot be otherwise, for to constitute types and assign designations that would include even the chief of these varieties would be to multiply both types and names beyond the capacity of the human mind to learn, or its power to retain. In the case of these diseases, therefore—diseases that are so frequent as to make up at least half the cases of affections of the nervous system which the practitioner has to treat,—the method by which he has acquired his knowledge, and by which he is able to deal successfully with almost all other classes of disease, fails him entirely; and yet in most instances he has nothing to put in its place. "The old order changes"; but it is "giving place to new" with a slowness which, at the present day, is a source of perplexity and difficulty widely felt, and constitutes a really serious obstacle to the practitioner's work. That it is so—that I am not exaggerating the facts—will be, I am convinced, confirmed by most of those who bring to their work the earnest and thorough spirit which strives to understand the cases that have to be treated, which strives after light, and is not content merely to deal, in darkness, random strokes.

What, then, is the remedy? What new method of diagnosis can be adopted? Happily, there is a means of meeting the difficulty, before which its formidable features will vanish, and the ever-growing accumulation of facts, the ever-increasing variety of definite morbid states of the nervous system, however numerous, however complex, however unfamiliar, may be dealt with as readily, as surely, as the diseases of any other class, or as the affections of the nervous system that do correspond to types, and may be recognised by such correspondence. Indeed, I had almost said that the method I am about to recommend enables you to deal with these diseases more readily and more surely than the other. At least, it has the advantage of placing the practitioner at once in the position needed for the adequate and wise treatment of the malady so far as such treatment is within our reach.

The method you should adopt is this: Whenever you find yourself in the presence of a case that is not at once and completely familiar to you in all its details, forget for the time all your types and all your names. Deal with the case as one that has never been seen before, and work it out as a new problem, *sui generis*, to be investigated as such. Observe each symptom carefully, and consider its significance. Then put the several symptoms together, and consider the meaning of their combination, especially whether there is any one part of the nervous system at which disease might cause them all. Lastly, consider the way they came on, as indicating the nature of the lesion, comparing this with the evidence of their seat, and remembering also that their character may in itself tell you something of their probable nature. When described in the abstract, this may seem a lengthy process. It may even seem a formidable process. As a rule, it is neither. The common symptoms, even those presented by uncommon cases, are not numerous, so far as concerns their general character and actual nature. The question of localisation is only an application of the common physiology of the nervous system, of the facts that should be familiar to every student, and can be re-learned, if necessary, with ease, by every practitioner. All the knowledge needed for this method is that which every student gains, or ought to gain, in the course of his studies; it is only the mode of using the knowledge that is new and has to be acquired. But the student should remember the great importance of "keeping up" his physiology and anatomy of the nervous system, or at least those parts of it which are needed in practical work. There is no department of medicine that consists more largely of applied physiology and applied anatomy than these diseases. For this reason they should engage the attention of the student early in his hospital work, instead of, as is often the case, being relegated to a late period on account of their supposed difficulty—a period when his science has got "rusty," or has slowly vanished, until even its nomenclature awakens a mere echo from bare walls. As a matter of fact, much of the student's hindrance is due to this postponement. The application of his knowledge should be made to retain it.

But for the successful use of this method it is essential that the knowledge, though neither extensive nor profound, should be firmly grasped and boldly used. Herein lies the chief practical difficulty. Timidity is almost a greater hindrance to diagnosis than is ignorance. You must feel sure of the meaning of symptoms, you must weigh the evidence with care, and then you may and must feel confident that your conclusion is trustworthy. This confidence and boldness can only be acquired by familiarity with the process, by observing its use by others, and afterwards repeating it for yourselves, thus becoming so familiar with the language of disease that you can read it with ease, can see at once the meaning of its words, and perceive with readiness the significance of its sentences. It is by affording an opportunity for this that the attendance on the practice, especially the out-patient practice, of special hospitals, is of peculiar value to the student, whether pre-graduate or post-graduate. A series of cases of the same class pass before him, in each of which he can observe the character of the symptoms and the process of diagnosis, and thus he gains, in a short time, a familiarity with the features of disease, and a quickness in perceiving their meaning, that he could not obtain in a long period of work at a general hospital, where such cases are few and far between, and the lesson of the one is more than half forgotten before another instance comes before him.

In all this method of dealing with unfamiliar maladies there is nothing that is not within easy reach of the average student or practitioner. If any difficulty is felt, it is only, as in so many subjects, the first step that costs a conscious effort. Once learned, the method becomes more easy with each repetition, and is acquired the more speedily if repeated under slightly varying conditions. Soon, its steps cease to be consciously felt, and in a short time even a student who has been fairly grounded in the elementary knowledge, becomes able to use the method with precision, and succeeds, in four-fifths of the cases that would otherwise baffle him, in arriving at a diagnosis in which a physician whose attention has been given for years chiefly to these diseases, can find little or nothing to change. Note, moreover, that the method indicated is that which the most experienced physician has himself to adopt in the case of a large proportion of the cases that come or are sent to him. So infinitely various are the morbid states of the nervous system, so diverse their manifestations, that a very large number of the cases seen are practically new, even to a man of the largest experience, and if asked he has to confess that he has not seen before any case precisely similar to, and often not one even approximately resembling, that before him. Yet, by employing this method, he is able to arrive without difficulty at a diagnosis as sure as when the malady is of a common type. The only difference between his work and that of the student is that the latter has to adopt the method with scrupulous care, in a larger proportion of the cases that he sees. He has to deal with a larger number as essentially new problems, to be worked out *de novo*. If he does this, as I have said, his errors will be few and rare, and only a small proportion of the cases will baffle his efforts. But never forget the essentials of the proceeding. Remember that you must for the time discard entirely your types and names. When you have made your diagnosis in the manner I have described, then, and then only, may you consider how far the case corresponds to a type and can be called by a familiar or unfamiliar name.

The last point suggests a common difficulty. The desire for a name is strangely strong in the case of the majority of patients. Unless their disease is designated, they go away unhappy, discontented, distrusting. "But you have not told me what is the matter with me" is their parting plaint. What are you to do? In a very large number of cases no recognised name can be given to the disease that does not involve more error than truth. A few patients are sufficiently sensible to endure the knowledge of the fact and to be content with it. In the case of most, the best plan is to give them a descriptive designation, and to write it down, that there may be no mistake, or it will come back to you some day in altogether altered form, so changed that you do not recognise your own production. The descriptive designation need not necessarily be short, although often it can be, and it then satisfies as much, provided the words are unintelligible. "Cephalic dysæsthesiæ" makes many a sufferer content. "Paroxysmal neuralgic pains, due to damage to the third and fourth posterior spinal roots, owing to a stationary lesion of the intervertebral articulations," delighted one good woman whom I lately saw, and who earnestly desired to know what was the matter with her, no doctor, she said, having as yet informed her. But always adhere strictly to fact in these descriptions, and avoid words that are apt to mislead because used in various ways. If you use the term "sclerosis," for instance, the chances are that the patient will look up the word in some medical dictionary, will identify his malady with insular sclerosis, read, forecast his future, and become utterly miserable.

To sum up in a few words the necessary change. Discard in the first instance all attempts to identify or to name, and try instead to read the malady, tracing the symptoms to the seat of their cause, and discerning the nature of the morbid process by their character and course. The method has the great practical advantage of taking you at once to the elements that should guide your treatment, and of enabling you to treat wisely a case the like of which you have never heard of, and a name for which you may not know. Mind, I do not say that you need adopt this process in every case of disease of the nervous system that comes before you. There are many that do belong to familiar types, cases that are characteristic and preceived at once to be such—cases of common chorea, for instance, of idiopathic epilepsy, many forms of hemiplegia, and the like. But the method is needed in all unfamiliar maladies, and in all cases even of familiar aspect that present some unusual feature; often, such a feature, in the end, is found to indicate that the case merely resembles superficially that to which it seemed to correspond, and is really quite different in nature.

How real and how extensive is the need for the change in method I have described can only be realised by those who have to correct the mistakes or supplement the deficiencies of those who first attempt to make a diagnosis of such cases as I have indicated. In many cases a practitioner candidly confesses his inability. I often receive a letter saying, "I cannot make out what is the nature of the disease; I have seen nothing like it before, and I cannot find its features described in any text-book." Yet the case, as a rule, readily yields to the application of the method I have described. More often, however, the practitioner has endeavored to make a diagnosis by the familiar system of types, and has, as it were, forced the case into the receptacle that seemed to correspond most nearly to it in form, but into which it did not really fit, and has given it a corresponding name, more or less erroneous, and often ludicrously wrong. I abstain from giving you instances; the statement of the fact will suffice, and examples are so numerous that some one would probably find that the cap fitted him with unpleasing precision.

I should like to have given you some instances of the working of this method, but for this there is not time. You will find it easy and more useful to work out such examples for yourselves, beginning with familiar diseases, but treating them, for the occasion, as unfamiliar. I hope, however, shortly to give you in another lecture a series of examples of its use, although not nominally such, by describing and analysing some of the more important diagnostic signs, or "diagnostic guide-posts" as we may call them. Familiarity with these is of extreme value, and most of them will afford illustrations of the use of the method I have just described.

LECTURE II.

MISTAKEN DIAGNOSIS.

Gentlemen :—It is always a pleasant thing to be right, but it is generally a much more useful thing to be wrong. If you are right, all that you do, as a rule, is to confirm your previous opinion, your previous habits of reasoning, and your previous self-esteem. But if you are wrong you generally gain in knowledge and gain perception of the way in which your method of diagnosis needs improvement, and the influence on self-esteem is not likely to do you harm. At least that is my own experience, and I think I have observed it confirmed in others. But the result is dependent on deliberate effort. There is a strong temptation to smooth down error, and it is very easy not to gain from it its precious lesson. It is more easy to fancy that there is some accidental cause for the mistake than frankly to perceive that it is a fault. But if you make a deliberate effort to realise and to face in your own mind the mistake you have made, to discern its cause, and to employ this perception as far as you can to remove the cause and prevent a like mistake in the future—if you do this, almost every error becomes one of the precious experiences of your practical life.

Yet you will note that I have not said this quite absolutely. I said it is "generally" more useful to be wrong, that "almost" every error is useful. As a matter of fact, there are errors you cannot avoid. Beware, however, of

Post-graduate Lecture, *British Medical Journal*, July, 1894.

thinking that any individual instance of error is of that character. The chances are great that it is not, that it might have been avoided, and they are considerable that it will be avoided in the future as a consequence of its occurrence. At least I hope so.

Still, there are cases in which error is inevitable. I should like to mention to you an instance which illustrates the conditions on which the relation depends. But this is not the chief subject I desire to bring before you. My chief object is to seize the opportunity afforded me by the courtesy of a private patient and his desire to be useful—the opportunity of illustrating avoidable mistakes by an example almost perfect in its character, almost unique in its features, and seldom equalled in the variety and importance of the instruction it affords. The case is indeed full of important lessons, and it is one the like of which, I venture to say, not one of you is likely to have seen before or is likely to see again.

But, first let me say a few words about inevitable errors. There is a patient in the hospital at this moment whom I cannot show you now, but whose case is an example of mistakes that must be. I saw him first about two months ago, at the commencement of his illness. I then made a diagnosis, of the correctness of which I felt assured, and which I am now certain was wrong. Yet if I saw a like case to-morrow with the same symptoms, with all the knowledge I have gained from this case, I should make the same diagnosis, that which proved to be wrong; and the chances are fifty to one, or more, that it would be right. Why, in this instance, was it wrong? Because all diagnosis that rests on reasoning is a matter of probability; only that which is simple observation is certain. The probability may be great, or may be only that of a slight preponderance of the balance of evidence, but wherever inference comes in there is no certainty. Remember that inference plays a part which you can discern only by an effort, in a large part of diagnosis. You think you observe the presence of consolidation of the lung or of pleural effusion. You do nothing of the kind. You infer its existence from certain physical signs. I remember an incident which illustrates the risk of error thus involved. A skilled physician gave orders to his house-physician that a surgeon should be called in to tap a chest, one side of which was full of fluid. When the surgeon came he took notice of the chest before plunging in his trocar, and, being a man of a thoroughly scientific mind and used to observation, he was struck by the fact that there was no enlargement of the chest. Although he was a surgeon, he percussed carefully, and found that every other sign of pleural effusion was present except enlargement of the side. He declined to tap. A little later the patient died, and the lung was found to be a solid mass of cancer. That was a case of incomplete observation, but it illustrates the fact that even a condition like pleural effusion, which you seem to observe, is really recognized by complex inference, and if either the process of reasoning or the observation is deficient, conclusion may be wrong.

To return now to the case I mentioned. The patient is a man of forty-eight years. I saw him a few hours after he had been seized with hemiplegia, which had come on in a few minutes. Such a sudden onset means a vascular lesion, the rupture, or closure of an artery. The onset had occurred without any loss of consciousness and without prodromata. There was no sign of degeneration or of kidney disease. The ophthalmoscopic appearances were normal. The hemiplegia was complete; but the entire absence of any loss of consciousness made it distinctly unlikely to be hemorrhage; and the facts that there was no tension in the pulse, and that there was a perfectly normal heart free from hypertrophy, made rupture of a vessel still more improbable. The absence, so soon after the onset, of any indication of a source of embolism excludes embolism—as a practical fact; and the absence of kidney disease made atheroma at that age most improbable. When no evidence of any other cause of arterial occlusion can be discovered, and there is no condition to cause the rare spontaneous thrombosis in a healthy vessel, syphilitic arterial disease is probable. It is the condition left by the exclusion of others, and *it* can very seldom be excluded. I came, therefore, to the conclusion that it was a case of syphilitic arterial disease leading to the sudden formation of clot in a branch coming off from the diseased part of the artery. But not only could syphilis not be excluded in this case; there was a history of it. The patient was perfectly conscious and able to talk, and gave me a history of syphilis more than twenty years ago; an intimate friend, moreover, told me of a further history of syphilis as recently as eight years ago. Under these circumstances the diagnosis of syphilitic arterial disease and thrombosis was the proper diagnosis, and treatment was arranged accordingly.

The patient was admitted here a short time afterwards, and, instead of improving as he should have done, he has got worse. By "improving" I mean in the general cerebral symptoms, not necessarily improvement in the hemiplegia. You cannot by treatment restore the lumen of the vessel or the tissue destroyed by necrotic softening. The treatment is capable only of influencing the condition of the wall, and not its consequence. Not only has the hemiplegia continued unchanged, as would be compatible with the diagnosis, but the general cerebral symptoms have become gravely worse; his mental state has become increasingly dull, weakness of the sixth nerves has developed, and, although the optic discs were perfectly normal when I saw him, optic neuritis has developed with a rapidity and to a degree of intensity that I think I have never seen before. It is certain that he is suffering from a tumour of the brain that is not syphilitic, and that in some way by pressing an artery the tumour led to the sudden formation of clot in it, and the sudden symptoms of the arrest of the blood supply in its region.

Of course, I need hardly add that he was treated by iodide of potassium in full doses, and soon afterwards mercury was added and pushed just to the degree of affecting the gums; yet not only without improvement, but with the contrary. I may incidentally add that the mercury was pushed to the degree of affecting the gums because, unless you see indications by the gums that there is enough mercury in the blood to act upon the tissues, the presumption is that the mercury is eliminated so fast that there is not enough in the blood to act upon the syphilitic tissue.

Such a combination of symptoms is seldom met with; but we do occasionally meet with a combination of symptoms which illustrates the same rule—that the diagnosis which it is right to make is wrong in fact.

I remember another instance of inevitable error in diagnosis, which I will briefly mention. An old woman, aged sixty, was admitted in profound stupor and with indications of bilateral hemiplegia and irregular damage to the cranial nerves. This state had come on so suddenly we had no doubt that there was occlusion of the basilar artery. She was sixty, and presented signs of degeneration: therefore thrombosis in a branch of an atheromatous artery was probable. But she had also mitral constriction, the cardiac lesion which of all others is most prone to cause embolism. Embolism of the basilar artery, curiously enough, has been thought to be impossible, because the artery is larger than either of those by which the blood reaches it, but manifestly a plug may pass through one of the vertebrals which cannot pass through either of the posterior cerebral arteries, which are smaller than the vertebrals. As a matter of fact, I have seen a distinct embolus in the anterior extremity of the basilar. Therefore embolism was also probable, for no age excludes it. The patient died, and was found to have syphilitic disease of the basilar, perfectly characteristic. This we had no reason to suspect. It is another instance of a case in which a correct diagnosis was impossible, and error was inevitable. From such cases only the general lesson can be learned, that accuracy is occasionally impossible; we can only be right in nineteen cases by being wrong in the twentieth. It is well to realize this. But remember that in practice we have to treat that which is only probable as if it were certain. We could not treat two-thirds of our cases properly without doing this. But always discern the degree of probability, and if the probability is low, reduce or modify your treatment so as to do as much good as you can, if your opinion is right, without certainly doing harm if your opinion is wrong. You will find that this principle is applicable to many and various contingencies.

And now we pass to the case which I want especially to make the subject of your attention to-day; I should rather say "the object," because the attention that I want you to give to it is to a considerable extent by the use of your own eyes. The patient is a man who first consulted me on February 9th, complaining of numbness in his hands and feet, with pains, and twitching of the muscles, which were increased when he walked. These symptoms were of three years' duration and had gradually become worse. They began after overwork. He had sharp momentary pains in both feet, sometimes in the toes, sometimes in the soles, and sometimes in the balls of the feet—not much in the legs, and none in the arms. There was no trace of knee-jerk; although there was a little complaint of unsteadiness, he could stand, with his eyes shut, fairly well. There was no inco-ordination in the hands, and no deficiency in their sensitiveness. The pupils acted to light. He had had slight brief double vision. There was diminution in sexual power but no difficulty in micturition. These symptoms, especially the loss of the knee-jerk and the sharp pains, afforded a strong presumption of tabes. Inquiry after the common antecedent of tabes gave no evidence of it, but did not enable it to be absolutely excluded. He came late one morning, after I had finished my work, and I had no time to examine the state of sensibility upon the legs; I made a note that it was to be examined next time I saw him. I generally refuse to see a patient unless there is adequate time for investigation, but it is not easy to resist the urgent desire for an interview when a patient has come from a distance. I made the diagnosis of probable tabes, and ordered him a mixture containing belladonna, quinine, and arsenic. He came again, about five weeks afterwards, saying that he was about the same. The pains were a little less, but the other symptoms were still troublesome. I then procceded to do what I had not been able to do on the first interview—to test his sensibility.

Among the many aphorisms I heard from the lips of the greatest bedside teacher whom any living person remembers, was one that flashed across my mind when, to test sensation, the patient's skin was bared. I remember hearing Sir William Jenner once say: "Gentlemen, more mistakes are made, many more, by not looking than by not knowing." To my astonishment, almost to my consternation, I saw that the skin presented everywhere the characteristic pigmentation produced by arsenic. It was a case of arsenical poisoning simulating tabes. And I had prescribed arsenic for him! Before his illness, and during its first year, he was by occupation an oil and colour merchant, handling papers of various tints, and all sorts of pigments, many no doubt containing arsenic. He had also during the first year he began to suffer taken a tonic mixture containing arsenic; but he did not take this long enough for it to do more than intensify the poisoning, which had no doubt been the result of his occupation. He had been exposed for many years; and it is probable that during the two years which elapsed between the development of his symptoms and the time I saw him, that which I did others had done. Arsenic had been given him to cure his symptoms.

I need hardly say that I changed the prescription. He has not, however, improved. I gave him iodide of potassium, which seems to have a definite action in eliminating arsenic from the system; but the course of his symptoms during the last three months is curious; they have rather increased than lessened, and especially the condition of the skin that I will show you in a moment, has become more intense. Unless there is some continued cause of arsenical poisoning, and we cannot discover it in any way, I think the iodide of potassium must have been eliminating the arsenic from the tissues to such an extent as to increase the amount in the blood to a degree that has further irritated the damaged structures.

Before, however, we examine the skin, let me remind you that the nerve symptoms in arsenical poisoning are most important. In acute poisoning they are met with after the acute symptoms have subsided; they come on gradually, and for a time increase. The arsenic taken in during the acute poisoning seems to enter into the nutrition of the nerve elements and gradually to derange their function and their structure. In chronic poisoning there is a gradual interference with function.

This fact, that arsenic passing into the nerve structures, perhaps partly in place of phosphorus, should first gradually modify their function, is not surprising. Function depends upon the release of force—nerve force—by chemical combinations; and the amount of nerve energy latent in complex compounds, released when similar compounds are formed, depends upon the constituents; so also does the readiness with which it is evolved and the kind of stimulus that excites the release. But the change in nutrition which at first disturbs function in greater degree, changes the structure, and leads even to disintegration of the substance of the nerve elements. As a fact, this is what arsenic and other metallic poisons do. Their action presents certain features which you should always bear in mind. They reach the structures by the blood. The structures on the two sides have bilateral symmetry in intimate constitution as in outward form, and that involves a similar susceptibility on the two sides. Hence it is that these poisons produce symptoms that are bilaterally symmetrical.

Moreover, they produce symptoms which are limited according to function. There we have a mystery, and yet it is a mystery we must recognise as an indication of fact. Function must be related to the chemical and molecular constitution of the nerve elements. The difference in susceptibility to different forms of force must depend on atomic constitution and molecular arrangement. We cannot conceive that the nerve endings in the skin which produce a nerve impulse on the slightest touch are quite the same in constitution as those which produce a nerve impulse when warmth is applied. We cannot believe that the same kind of structure can subserve the susceptibility to the massive motion of a touch or to the wave motion, only a little less frequent than the waves of light, which constitutes heat. And that difference, we can understand, is equal or greater in the motor and sensory terminal structures on which toxic influences for the most part primarily act. This enables us to understand why it is that poisons act specially upon structures with certain functions, and should even act first on one set of sensory nerves and not on another set. So we have, from different poisons, different effects upon the nerves, and we have also different effects from the same poison, probably dependent on the fact that it enters the blood in somewhat different form, or with greater or less rapidity, and thus is assimilated by and deranges certain structures rather than others. Thus we have, from arsenic, palsy of the extensors of the arms and feet, as we have from alcohol. We have sometimes a perfect simulation of tabes, ataxy, muscular anæsthesia, and loss of knee jerk, in consequence of the preponderant affection of the afferent muscle nerves, and sometimes we have chiefly sensory cutaneous symptoms, although scarcely ever without loss of the knee-jerk. Remember, that loss of the knee-jerk only indicates affection of the afferent muscle nerves, and that we know that the impairment of these nerves has many causes. To all toxic influences they seem specially susceptible.

Hence, I beg you to remember, above all things, as a practical rule, that whenever you have bilateral disturbance of the peripheral nerves, symmetrical, with a distinct limitation to function, sensory or motor, it is almost certainly due to a blood state. There is reason to believe that the degeneration which may appear to constitute an exception is really due to a blood poison. Many are the cases in which the recognition of that fact will save from grave error.

In the nerve symptoms of this patient there is little more for you to observe than I have told you, but if you ascertain for yourselves the absence of the knee-jerk, the fact may be more securely fixed in your mind, and it cannot be too surely held. But, as regards the cutaneous symptoms, there is a picture for you to study such as you will not see again. This arsenical pigmentation is so characteristic that the moment I saw it I was sure of its nature, but I never like my own certainty to be without corroboration if it is outside my own special region. So I sent the case to Dr. Radcliffe Crocker, and he is also absolutely certain. He has seen, he said, one more intense case, and he has kindly allowed me to show you a plate from his *Atlas*, which illustrates it.

I will also read to you his description of the pigmentation. But I will add first one more remark. How I have become familiar with the pigmentation from arsenic is because I have often seen it in the cases of epilepsy. I have also seen it in well-marked form, in a lady whose love for working on muslins led her to their continuous handling, and she preferred æsthetic tints. The result was that she began to suffer from pains that were ascribed to gout, until there came progressive palsy of the extensor muscles of arms and feet. This showed a toxic influence, by the symmetry and limitation I have mentioned. Then the pigmentation was discerned, although, I believe, chemical analysis had already made certain the cause. But the troublesome pustular eruption which bromide causes in most persons can only be prevented, or kept down to insignificant degree, by arsenic. No other drug, no modification of the way or form in which bromide is given, has any influence. Some patients are peculiarly prone to the rash, and have to take the drug for a long time; they have to take arsenic also. A slight degree of pigmentation is often produced in such cases. It has, therefore, several times happened to me that I have had to put to the patient the two evils, and say, " Which will you have, the spots or the pigmentation?" There is no other help for it; they must have one or the other. They always choose the pigmentation. It does not, however, increase to a really serious degree, and I have not met with any other symptoms of chronic arsenical poisoning.

If you observe carefully the skin of this patient, you will

see that the pigmentation begins as small dots, rounded, which coalesce, but leave small rounded areas of unpigmented skin. These appear whiter than normal, perhaps only by contrast. The pigmentation is abundant everywhere, but extreme on the neck, front and back of the trunk, arms and thighs. For the most part it appears as a pure process, but in many places you will observe small round spots of congestion, of the same size, which suggest that, at least when the process is going on actively, congestion may be the first part of the process. Moreover, in the neck, where the skin has been subjected to habitual friction, the pigmentation is so intense as to be practically continuous, and with it so much congestion is combined as to give the skin a reddish or purplish-brown aspect. The congestion is also conspicuous on the hands, and there, on the palms especially, the spots are attended with distinct elevation. Indeed, where the epidermis is thick, there are minute raised elevations, at which no indication of congestion or pigmentation can be discerned. The change in the palms, as you have just learned from Dr. Radcliffe Crocker's description, may go on to a form of "palmar psoriasis."

The great fact is the pigmentation, at first essentially punctate or in small spots. It is that that is so characteristic. You will at once understand how important it is to be able at once to recognise a sign so distinctive and so significant. You cannot note the state too carefully. Once firmly fixed in the visual memory, it will not be lost. It is seldom easy to retain an image securely from a single opportunity of inspection, but this condition is one, I think, which you are not likely to mistake when you see it again. You may have some other opportunity of observing a similar condition. You probably will have such if you carefully examine the skin of patients who have been, for long, taking arsenic with bromide. But the story of the origin of this patient's illness suggests that a careful inspection of the skin of a number of men employed in shops where paints and papers are sold might afford an inquiring student opportunities of observing the pigmentation, which would be beneficial not to himself alone. It is certain that the change in the skin may long precede any other symptom.

Having helped you so far, gentlemen, I must now ask you to help yourselves. Observe, also, what I have not mentioned, the bilateral symmetry of the rash, a feature that you will recognise, from what I have said, to be inevitable. Connect it with the bilateral pains and loss of knee-jerk, and fix these features of limitation and symmetry in your minds in association with the toxic cause. For a moment, indeed, forget what the cause here is, forget the precise character of the symptoms, in order that the general facts of combined functional limitation and bilateral symmetry may stand out more clearly in your view as landmarks which, when and wherever they are seen, will always guide you surely.

LECTURE III.

ARGYRIA AND SYPHILIS.

Gentlemen:—It is the duty of a clinical teacher to bring out of his treasury things old and new. He is constantly under some temptation to present ideas that are new, because they possess more intrinsic interest, although it is generally more useful to give those that are old. Truth and novelty are by no means necessarily associated, although a familiar phrase which suggests mutual exclusiveness goes too far. Nevertheless, it is well for a teacher, if he can, to resist the attraction of the new, and it is always unwise for him to hesitate to inculcate that which is old merely because it is old. Years ago, when I was engaged in giving the elements of clinical instruction to students who were beginning their practical studies, I used to paraphrase the saying of Demosthenes regarding oratory and maintain that the first thing in learning is repetition, the second repetition, and the third repetition. A teacher should remember that to neglect to repeat is an unpardonable sin. There is an unpardonable sin of the student; in fact, it was the remembrance of this that made me use the term. It is not my own idea. Years ago Sir William Jenner used to say to us (and whatever Sir William Jenner said may be *a priori* regared as certain) that the unpardonable sin of the student was to say "Yes" when he ought to say "No"—to say that he heard a thing, that he felt a thing, that he understood a thing—when he did not.

For you this lesson is probably useless, because you have

all passed the stage at which it is needed. But it may not be too late to remember that it is a mistake to shrink from unattractive repetition. Consider—what are we all learning, or should be learning, on this point, from our great Teacher? What does Disease impress upon us? Disease is forever repeating to us the same things. All the more important laws and rules are ever being pressed upon us, in varied tone, by varied emphasis, or in varying language, but ever repeated. No repetition should be or will be useless to us if we take the requisite pains, and are not deterred by weariness from striving to discern the lessons that seem the same, but present always an important difference; and no repetition of fact to us by disease is ever superfluous, unless we will to make it so.

Yet the teacher is, as a rule, compelled to take as a subject for his instruction some fresh illustration of disease. It is seldom that he has an opportunity of taking that which is old. When he can it is well for him to do so, for more than one reason. All old cases afford an opportunity of observing the changes in disease, and the effects that time permits. The question which ever presses on us when we meet with disease in its active stage is this: What will be the future; what will be the result? We can only learn to look forward by taking every opportunity of looking back. We cannot combine in our observation the future and the present. Our own experience and that of others may enable us to guess something of the future from the present; but every personal observation that increases our ability to forecast, everything that helps to make the guess more than a guess, is important. This help can only be obtained when the future has changed to the present. We can realise the past, and discern its relation to the future when that which was future has come. We cannot, with the same confidence, realise the unknown future and apply it to the present. Some of you may remember the lines which come to my mind—lines by a poet who is almost forgotten now, although his name should live in the memory of medical readers. Sidney Dobell wrote in one of his sonnets, in which sadness touches softly—

> "And when the now is then,
> And when the then is now."

If the "then" is in the past, we can, by fancy, make it "now;" the "then" is clearly seen, for it is now a fact, and we can thus gain some secure experience in prognosis.

I show you to-day a patient who presents an opportunity of illustrating the present by the past, and by a past in itself most instructive, such as seldom occurs to a teacher. I imagine it is almost a unique opportunity for a teacher to be able to take as a subject a patient who has been under his observation for a quarter of a century. That is the case here, and I am glad to have so unusual an opportunity in beginning again the series of our Wednesday lectures. It is nearly, if not quite, twenty-five years since this patient was under treatment in the wards of the hospital in the acute stage of his affection.

Did you notice him as he came into the room? If you did not, you should have done so. One of the habits to be acquired, and never omitted, is to observe a patient as he enters the room; to note his aspect and his gait. If you did so, you would have seen that he seemed lame, and you may have been struck by that which must strike you now— an unusual tint of his face. Those two things are important. They are, indeed, connected, but in a way that is rather curious than instructive. It is, indeed, so curious that I cannot resist the temptation of telling you the story it involves.

The patient came here in 1870, with symptoms of a cerebral tumour, of rather rapid onset for such a morbid process. The symptoms had reached to a considerable degree in about two months. The patient presented indications of a sub-chronic local cerebral lesion, with headache and optic neuritis. These two general cerebral symptoms with the onset indicate that the local process is a growth. Moreover, there was a history of active syphilis; and we know that whenever we have evidence of a local growth of rather rapid course in the subject of syphilis the probability is very great that the growth is syphilitic. They are much less if the growth is very chronic, and this point is important.

The patient was treated according to the diagnosis. He was not under my care, although he was under my constant observation; I was then Medical Registrar to the Hospital, and he was under the care of Dr. Hughlings Jackson, from whom it was my privilege to learn many lessons of ever-increasing value, and not the least important were connected with this case. After the patient's discharge from the hospital he was under my care as an out-patient, of late years only seen occasionally, chiefly for the benefit he is always ready to give to others as an illustration.

When he first came to the hospital he was lame, as he is now, and he presented the complexion aspect you see, but in a greater degree. An inquiry into his history showed that two years previously he had been an in-patient at a general hospital, under the care of a physician then well known, who has now been dead many years. The symptoms he then presented were those of a small syphilitic growth pressing on one side of the spinal cord, and causing effects that we now know to be very characteristic. For those symptoms he was treated with nitrate of silver, and his skin acquired the aspect which it has never lost. When he came here he had improved, and I think, for several reasons: it is exceedingly likely that after the nitrate of silver had been given for a considerable time, without other result, mercury was substituted. At any rate, the affection of the leg ceased to increase, improved somewhat, and then became stationary, and when he came to this hospital its state was much as it is now, with due allowance for the effect of the fresh trouble which brought him to us.

In connection with his case there is a little story which I cannot resist the temptation of telling you, especially since the patient here can correct me if I am wrong. Perhaps it is a little beyond the proper subject of a lecture, but I dare say, gentlemen, you will not be strict. The patient suffered from severe headache, optic neuritis, and signs of a local cerebral lesion in one cerebral hemisphere, of subacute onset. It was certainly a quickly growing tumor, and almost certainly syphilitic. Thus the case was most instructive. In 1870, as you know, not quite so much was known of optic neuritis as is now known, and he was shown to a good many visitors. The interest, too, was not lessened by the indications of argyria—the staining of the face from nitrate of silver for a morbid process in the spinal cord similar to that which in the brain lessened with extreme rapidity under iodide of potassium.

Great care and caution were taken in all that was said in his presence. I can even now remember the scrupulous circumlocutory care adopted to guard against any perception, on his part, of what was thought about his previous treatment. But the man possesses a considerable amount of intelligence, and he picked up too much information, although he gave us no indication of the fact. The symptoms rapidly subsided. On the morning after his discharge he paid a visit to the physician under whose care he had been, and from whom he had received the silver. He obtained an interview with the doctor. The result of that visit was, I am certain, to improve the therapeutical knowledge of the physician, and I have also no doubt that the result was very much to the advantage of any other patient who subsequently came under the care of that physician for a similar affection. But the immediate result was a considerable disturbance of equanimity. The patient was wise enough to content himself with thus conveying instruction. He might, I think, have gone further; but I doubt whether even a speculative lawyer would have induced him to do so, for he is, after all, a reasonable fellow. I think one cannot find very much fault with the lesson, or even, considering all things, with the way in which it was given.

[Dr. Gowers here turned to the patient and asked if the account given was substantially correct. The reply was: "Yes, sir; it's all right. I *jacketed* him."]

I imagine that it is very likely that some of you have never seen the tint of argyria. It is less commonly met with now than formerly, because nitrate of silver is less frequently given, and when it is given, it is given with more care. It will therefore be wise for you to note very carefully the aspect of this patient. The tint is rather less than it was, but it persists, and it will persist as long as he lives. There is not now a black line at the edge of the gums; I think it was there formerly. We have been unable to find the old notes of the case, but all the essential facts are adequately impressed upon my memory. I have only myself seen about four cases of staining from nitrate of silver. This is one. Two were in cases of epileptics, for which it was, as you know, once a reputed specific. It was held in very high esteem by some persons in what have been called the "pre-bromidic days." Both the patients I saw, who had been stained with nitrate of silver for epilepsy, were still patients here for the persisting disease, and therefore my own observation did not lead me to entertain a very high opinion of its value.

The fourth case is instructive because it was due to the use of nitrate of silver for the good it can unquestionably do in gastric affections, especially when pain occurs before meals, that is, when it coincides with the absence of food. Although in cocaine we have an agent which seldom fails under these conditions, it is not unlikely that this use of silver will again increase. It may be then that the lesson this case gives may be again needed. A doctor, an esteemed practitioner in a suburb of London, gave his brother, who suffered much gastric pain, a "dinner pill," to be taken before food every day; it contained some oxide of silver. A year or two after, when shooting together in Scotland, the doctor became uneasy at his brother's cyanotic aspect. He watched him closely, and at last asked him: "Do not you get short of breath as you go up hill?" But there was no shortness of breath, and the doctor did not think anything more about it. Six months later, the tint had increased, and it suddenly flashed across the doctor's mind, "Why, it must be silver; there is that pill!" He turned to his brother and said, "Have you been taking those pills ever since I first gave them?" "Yes," was the answer, "I have been taking them ever since, and am still." By this time his face had become deeply tinted. The line on the gums was most distinct. Although they were at once stopped, other remarkable troubles came on. There were no signs of lead poisoning, no colic, and no conceivable source of saturnism, but the patient developed wrist-drop, just like the wrist-drop of lead poisoning, and also developed the gout that is so often due to lead, and he developed the albuminuria associated with it. Silver does cause palsy in animals; we know how many metals may cause the symmetrical extensor palsy, and I think there can be no doubt that in this case the palsy of the extensors of the arms was due to the silver, although the case, as far as I know, in that respect, is unique. The sequel of the symptom, I may add incidentally, was most illogical. The patient died two years later, but he died from cancer. With adequate mischief to terminate life by intelligible effects, he died from something altogether different, which is an illustration of the limits there are to our power of inference and forecast.

Let us turn to the patient before us. He left the hospital with some symptoms remaining from the disease in the brain; these had become stationary, and they have persisted ever since. I said he had symptoms of a local cerebral lesion. These symptoms were slight left-sided weakness, left hemianæsthesia to all forms of sensation up to the middle line, head, limbs, and trunk; considerable diminution of the special senses on that side—taste and hearing, while vision was affected as hemianopia. Of smell I am not sure. If impaired, the defect did not persist, and his present recollection is that it was normal. The hemianopia was at first complete.

Leaving smell out of consideration for the moment, you know that left-side hemianæsthesia, involving the skin and the special sensories, is the characteristic of what is called hysterical hemianæsthesia, a functional condition of the existence and reality of which there can be no doubt. Nor can there be any doubt of its practical independence of the patient's will. There was the difference, however, in this condition from the hysterical form—that in the latter the affection of vision is a diminution in the whole field of vision. There is a considerable general diminution in the field of the opposite eye, and a slighter similar diminution in the field on the other side—that is, the side of the central hemisphere involved. In this case, however, there was hemianopia on the opposite side. That is the great difference between this form of hemianæsthesia, that which is due to organic disease, and that due to a functional affection. Nevertheless, there is evidence to show that " crossed amblyopia," dimness of vision, with general restriction of the field, greater on the opposite or "crossed" side to the affected hemisphere, may occur from organic disease as it does from hysteria. But the reason why an organic lesion generally causes hemianopia is this. The lesion is generally at one place, at the region which Charcot has called the "sensory crossway," at the posterior extremity of the internal capsule between the optic thalamus and the end of the lenticular nucleus.

The sensory fibres from the skin, from the head, and limbs run in the posterior third of the hinder limb of the capsule. The optic tract conveys impressions to that region, the fibres from the same-named half of each retina conducting impressions from the opposite half of each field of vision pass to it. Thus the impressions passing to this hemisphere are those from the side on which the motor processes act on the limb. Each half of the brain receives impressions from the side on which it moves the limbs.

The fibres from the optic tract that subserve vision pass into the white substance of the occipital lobe. They probably have an intermediate station in the posterior extremity of the optic thalamus. But I need not now dwell on this. Whether they pass from the thalamus or directly from the tract they must pass close to the extremity of the capsule, close to the sensory fibres from the skin. Thus disease here causes hemianopia and cutaneous anæsthesia. But it also causes loss of taste and hearing. This is so well established that we are sure that the paths for these sensory impressions pass by this region. Moreover, there are cases on record of an affection of smell from disease of this region, but the point needs further evidence, and as this case has no definite bearing on it, I will pass it by. But I must emphasise the fact that cases of organic disease have been met with which cause symptoms resembling closely the hysterical form of hemianæsthesia. The latter we must ascribe to an inhibition of the sensory structures in one hemisphere. In the cases of organic disease that cause similar symptoms, extensive disease has existed on the convexity of the hemisphere, so extensive that we cannot infer more than that it is in the convexity of each hemisphere that the impressions are represented which come from all the special senses of the opposite side, including smell. In these cases there is "crossed amblyopia." But, beyond recognising the fact, you need not now consider the condition. Our subject is the sensory effect of disease that causes vision to suffer, as "hemianopia." Half-sight is lost with the other senses if the disease is in the tract, the sensory-crossway, or the half-vision centre in the occipital lobe. This was the combination the patient presented. The association of hemianopia and impairment of the general and special senses on the same side proved that the disease was situated where all the paths are in contiguity—that is, at the place I have mentioned. Beyond this the paths diverge, so that the combination can be produced only by disease invading the whole cortex of the convex surface, including the occipital lobe, and then there is the crossed amblyopia as well as the hemianopia. There is restriction of the field as well as half loss. We may now turn to the symptoms the patient still presents, the permanent residue of those caused by the active disease. Persisting as it has for the past quarter of a century, we may actually expect it to persist for the next quarter of a century, which is probably about as long as the patient may be expected to live. A slight defect of taste and of hearing is still distinct. In the limbs there is still imperfect sensation.

The defect in vision that still persists is particularly instructive. You can easily verify it for yourselves. In one pair of charts before you, you see first the fields of vision, as they were nine and a-half years ago—that is to say, about fifteen years after he came for treatment—and the others present the condition that exists at present. The two are practically identical. In each there is a loss of the left lower quadrant. At first the whole half field was lost, up to the middle line; but the upper part recovered slowly, so as to leave the loss confined to the area you see. In it there are points that it will be instructive to you to note. The loss stops short of the fixing point. That, as you may know, is a characteristic of all forms of hemianopia, at any rate of lateral hemianopia. I show you another diagram, in which the loss is of the whole half field, in which the feature is well shown. The blind area reaches the middle line, above and also below; but the dividing line between sight and blindness curves round the fixing point, so as to leave an area of vision. This is the rule—I believe invariable. Around the fixing point there is vision on the blind as well as the seeing side. I believe that apparent exceptions are due to imperfect observation—imperfect almost of necessity, because the area of sight around the fixing point varies, and special means are needed to ascertain it when it is small in extent. The explanation of it is that from the region just around the macula lutea fibres pass by each optic tract to each hemisphere. Hence, disease of either optic tract, while causing hemianopia, does not cause loss in that small region from which the fibres go to each tract. It may have occurred to you that if fibres pass to each optic tract, disease of either should lessen the function of the whole of the region, although causing no absolute loss on either side. It is so. If you test minutely the central area of vision of a patient with hemianopia you will find that, although the region round about the fixing point is spared, the acuity of vision in it is definitely reduced.

Note also another point of significance. There is but little restriction in the general field of vision. The very slight diminution in the general area of the peripheral parts of the fields is not greater than can be accounted for by a dark day, or even by an individual variation; yet note how different is the extent of the remaining half fields in the other chart I show you. Notice the remarkable restriction of field, and that it is much greater in the eye on the side of the half loss—that is, the right half of each field being lost, the remaining part of the right field is much smaller than the left. I must not, however, allude to this to-day beyond asking you to remember it in connection with "crossed amblyopia."

And now, gentlemen, in conclusion, I have to impress upon you one practical lesson which this patient gives us, and which he gives as an old patient. It is the persistence, in some degree, of the effects of syphilitic disease, and that in spite of the fact that the patient was treated thoroughly, a few weeks after the development of the symptoms—which is as promptly as most patients are treated. It was manifestly adequate, for the urgent general symptoms, headache and optic neuritis, began to lessen within a week, and in the course of a few weeks all the acute symptoms had passed away, and the local symptoms were rapidly improving, the loss of power had become trifling, the hemianæsthesia gradually became partial instead of complete. Yet some of these symptoms have never quite passed away.

The optic neuritis when he came in was considerable, but not extreme. The whole disc was obscured by a swelling of moderate prominence and considerable vascularity, yet the acuity of vision was perfect. The inflammation was not sufficient in itself to have involved the nerve fibres so as to impair sight by the process of inflammation in them, nor was it sufficient to produce new inflammatory products sufficient to damage the fibres by their cicatricial contraction. Those are the two ways in which sight suffers from the neuritis. In those days ophthalmic surgeons generally refused to believe that considerable neuritis could exist with perfect vision. The scepticism was chiefly dispelled by Dr. Hughlings Jackson, and I think that this patient was one of the cases by which the dissipation was effected. Fortunately the treatment was early enough to remove the neuritis before grave damage was done to the fibres. His sight has remained good so far as acuity of vision is concerned, and his optic discs now present such a perfectly normal aspect that you would not suspect they had ever been inflamed. I should advise you to examine them carefully, because it is very seldom that you have an opportunity of seeing optic discs which are known to have been the seat of inflammation, and which now present no indication of it, and especially discs that were inflamed long ago.

The practical lesson that I mentioned, and which I should like you to take away, is this: The idea is not yet extinct that all syphilitic disease will yield to treatment, and that if only the symptoms are certainly due to syphilis they can be cured. Probably you know well as regards the nervous system how erroneous that idea is. You know that a syphilitic ulcer of the skin will, and must, leave a scar which nothing can remove. It destroys the tissue of the skin, and that tissue of the skin being destroyed is never replaced by structure which is like the old skin in aspect and function. So it is with the brain. If there is absolute destruction of tissue by a syphilitic process, that tissue cannot be renewed, and is not renewed; the symptoms dependent on its destruction will not pass away unless other parts of the brain can compensate for their loss. Remember that, as I have often said, the symptoms of a local lesion are never due directly to the syphilitic process. In true syphilitic affections, those which can be removed by iodide of potassium and by mercury, the syphilitic process is altogether outside the nerve elements themselves. These suffer secondarily, as they would from any other process of a similar general character. They suffer from compression by such a syphilitic gumma as this patient had, as they would from any other tumour. They suffer somewhat from the inflammation adjacent to any rapid growth, but this is usually simple inflammation with no necessary specific element. Through these processes they undergo damage and destruction, and no removal of the growth can do more than permit the recovery of those structures which are so little damaged that their recovery is possible. In the patient before you the damaged structures recovered. Destruction, of necessity, persisted, and the effects of the loss of tissue remain, where no compensation could be effected. That is the case with the half-vision centre and the fibres that carry impulses to it. Destruction of these causes destruction of the function and lasting loss. We have in the quadrantic loss of the fields of vision a proof of this fact. We have in it also evidence of the equally important fact that the effects of syphilis are often far more than that to which the term "syphilitic" can be applied. If you learn from this to look at a process with the imagination, that is as important in practice as in science; if you learn to discern the elements on which symptoms depend, and to be cautious in your prognosis in such cases of syphilitic disease; if you learn also the lessons the patient teaches regarding silver as well as syphilis, you will not have wasted your hour here to-day.

LECTURE IV.

SYPHILITIC HEMIPLEGIA.

Gentlemen:—You have probably long since become conversant with the truth that the most important part of learning is repetition. If this be true in the process of acquiring knowledge, it can hardly be otherwise in the process of imparting it. In conveying, as well as in receiving, instruction it is essential that the knowledge which has to be securely retained should be impressed upon the learner as often as opportunities permit. It may not be the most attractive element in the process, but another lesson which you have doubtless by this time learnt only too well is that the attractive is seldom the most useful, or the useful the most attractive.

Last week we examined an example of a rare disease and considered what is known of its nature, the process of its diagnosis, and the possibilities of its treatment. You had to hear of much that is conjectured but not proved, and of much that is still mysterious. To-day I propose to show you a patient who is suffering from a common malady whose symptoms can be traced to their source with reasonable certainty by means of the methods which must often have come under your notice, but which you will have to apply so frequently that they can scarcely be too often reconsidered. This is an advantage, not only for the sake of the repetition, but because no two cases of disease present precisely the same features in state and history; and in every case you will meet with one or more elements with which in that combination you are not familiar, and only experience in applying the processes of reasoning to the facts observed will enable you to place them in their proper relation and make them yield you the guidance you require.

The patient before you is suffering from hemiplegia, from paralysis of the right side. Common as hemiplegia is, one feature of this case should at once strike you as somewhat unusual. The forms of hemiplegia with which you are chiefly familiar have been in persons of advanced life, but this patient is only twenty-five years of age, and the affection has only existed for two months. Probably you have seen some cases of the affection in young persons, but not many in whom it has come on at this period of life.

If you note the power of movement which he possesses, you will see that he can move the upper part of the face as well on the right side as on the left, but he cannot move the lower part of the face so well. His tongue is not affected; it is protruded straight. His arm is feeble; he can move all parts of it, but cannot exert force with any part, and the movements of the hand are not quite steady. His leg is also weak, so that he can only just manage to walk alone.

When you are accustomed to cases of the kind, your first thought will probably be, and, indeed, may reasonably be, What is the nature of the lesion? and your second, What is its place? This is to put the chief elements in the diagnosis in the relative position which they occupy as regards the practical questions of prognosis and treatment as regards the interest of the patient. But the strictly logical order is to consider first where the lesion is, and, secondly, what is its nature. It is best to adopt this order in the case of diseases of the spinal cord, and we will adopt it now; but in the case of diseases of the brain there is no real inconvenience in considering first the nature of the affection.

The first diagnostic indications should be familiar to you. Paralysis of the face, arm, and leg on the same side must be the result of a lesion above the middle of the pons. Here the motor path for the face leaves that for the limbs and crosses over the middle line, and in disease below this point the face escapes or is affected on the side opposite to the limbs. Disease in the upper half of the pons often involves the fifth nerve on the side of the lesion,—that is, the side opposite to the paralysis of the limbs. Very rarely it does not, and the palsy resembles that which is due to disease higher up the motor path; but such exceptional cases, which are not met with in more than one case in a hundred, may be put out of consideration. Disease of the crus involves the third nerve on the side of the lesion, and opposite to the paralysis of the limbs. When these indications of involvement of the cranial nerves are absent you may safely conclude that the lesion is within the cerebral hemisphere. But where? It must be in some place in which it can damage the fibres that conduct the motor impulses, or the centres from which the fibres proceed and in which the impulses are generated. Experience shows us that there are three regions in which lesions are prone to occur and in which they may cause this effect. One is the central ganglia, the second the white substance, the third the gray cortex.

Lesions in the white substance of the hemisphere are not common. Moreover, if they are near the cortex the effects they produce are commonly very similar to those of disease which is in the cortex. If, on the other hand, they are near the central ganglia, the symptoms which they cause are like those that are produced by disease of the latter. Hence, for practical purposes, we may leave them out of consideration, and, indeed, we are compelled to do so, except in rare cases in which the diagnosis depends upon indications too complex to be included in a general outline of the subject, such as I am now attempting to give you. Thus our problem is limited to the distinction of disease in the two remaining situations. I have spoken of the central ganglia, but a lesion there does not cause hemiplegia by its effect on the ganglia themselves. Even extensive destruction of their gray matter produces no paralysis. The hemiplegia depends upon the interference with the narrow tract of white fibres which passes between them, and which, bounding as it does the lenticular nucleus on the inner side, is spoken of as the internal capsule. When the middle of this and the part behind the middle are diseased, hemiplegia is produced, for the fibres here are those which conduct the voluntary impulses from the cortex from its " motor " or " central " region. This, as you well know, is the part whence the motor impulses leave the cortex, and disease in this region causes hemiplegia like that which is produced by disease of the internal capsule.

How can we tell in which of these two situations the lesion is? First, by the fact that cortical disease means damage to nerve-cells, and that damage to motor nerve-cells is apt to be attended by the spontaneous action which is manifested by convulsion, while disease of the central ganglia does not cause convulsion. Secondly, from the fact that the close approximation of the fibres for the face, arm, and leg in the internal capsule and the disassociation of the centres in the cortex make it common for all parts of one side to be affected in capsular, and for only some parts of the side to suffer in cortical, disease. But this is not an absolute rule. All parts of the motor region may be affected in the cortex, and only some parts of the motor part in the capsule. But it must be a very small lesion in the latter to produce only a partial effect, and a severe lesion in the former to produce a complete effect. Here we have the indication of a guiding distinction. You must consider the significance of the distribution of the palsy in connection with its extent. In this case all parts of the face, arm, and leg have suffered, but only in moderate degree. Hence we may assume that the lesion was not severe, and the distribution, therefore, is strongly in favor of the lesion being in the region of the central ganglia. The tongue, indeed, has escaped, but this is not surprising if, as we perceive is probable, the lesion is incomplete. The diagnosis is further supported by the fact that the patient did not suffer from convulsions at the onset, and has not suffered from them since. If such convulsions had occurred, or if there were absolute paralysis of the arm and no affection of the face, we should be justified in concluding that the disease was in the cortex. You may thus perceive the chief points that must be taken into consideration in this point of localisation, and you will find that they will afford trustworthy guidance in a majority of the cases which come under your notice.

We may accordingly conclude with reasonable confidence that the symptoms are due to a lesion in the central nuclei, which has caused the hemiplegia by the damage, extensive but moderate in degree, to the fibres of the internal capsule. In scanning the brain of persons who have suffered from transient hemiplegia some time before, you may sometimes see nerve-fibres passing almost intact through an area of disease which involves the gray matter on each side of them, although the gray tint of the fibres on the surface of the layer shows that many of them have suffered structural damage, although there may not have been actual interruption.

The next question is, What is the nature of the disease? This is, indeed, a double question, for we have to consider first what is the lesion which has caused the symptoms, and next on what pathological process does that lesion depend. Several features of the disease are commonly available for ascertaining the nature of the lesion, but of these one exceeds all the rest in its importance in the readiness with which it is ascertained and with which it is applied, and the security of the information which it affords. It is the mode of onset. Each morbid process develops in a certain time, which varies, it is true, but varies only between limits, and these limits constitute our guide. The onset of the symptoms is that which manifests the rate of development of the process. For practical purposes, within which the majority of the cases will be found to come, three different modes of onset are to be recognised,—the chronic, the acute, the sudden. I need not amplify the words. A chronic onset means a tumor or sclerosis, an acute onset inflammation, a sudden onset a vascular lesion. Fix, I beg you, the last axiom firmly in your minds. You may rely upon it absolutely, and you will find the value of its guidance, not seldom, when you are perplexed by the intricate maze of discordant indications. Of course, I am speaking only of organic disease, and of these it is always true that a sudden onset means a vascular lesion. It means rupture causing hemorrhage, or obstruction causing, first, anæmia, and, unless at once compensated by a collateral flow, necrotic softening.

What was the mode of onset, what was the time occupied by the onset, of the symptoms from which this patient is suffering? If you put the question to him he will tell you that they came on in the course of five or six days. Such a period is that of an "acute" onset; it is the period which suggests that the morbid process is inflammatory. But you must never be content with general statements. Always strive to obtain details, even when they seem superfluous. How important the details may be is usefully shown in this case: they change entirely the significance of the first statement. We learn, first, that three weeks before the symptoms came on he had a brief attack very much like that with which the paralysis began. He is fond, as many wise men have been fond, of finding recreation in the violin. One day he found, while playing, that he could not draw the bow across the strings. It seemed as if it were glued to the strings. Then he found the right arm and hand were altogether weak. But in an hour or two all this had passed away. Three weeks later the same disability recurred under the same conditions, and it was accompanied by a peculiar feeling of numbness in the right arm and the right leg. When he walked he staggered, and so he continued until he went to bed. Next morning he found the arm and leg had become weaker. No change in them occurred during the following day, but in the next night complete loss of power came on, so that when he awoke he could not move either arm or leg. His speech was "thick," and, indeed, had been so upon the preceding day. Thus, when closely examined, we have a series of sudden attacks, the first in the day, after a sudden premonition such as had occurred three weeks before, the others in the night. The suddenness of onset during sleep we have to infer, but in most cases, when symptoms of considerable degree come on during sleep, their onset belongs to the class which we call sudden. We have further evidence of the truth of this in the fact that the first commencement was definitely, absolutely sudden, and succeeded a slight brief attack which was likewise sudden. So the details of the development of the symptoms show that the onset must be placed among those which indicate a vascular lesion. Which of the two processes had occurred? Cerebral hemorrhage is excluded by the age of the patient. It is not, indeed, excluded by the patient's age, in the abstract. Many of you may have seen cerebral hemorrhage in a person as young as he is. But you have seen it as the subject of a pathological, and not a clinical, demonstration. Note the difference implied in this distinction. Hemorrhage occurs in the young, but it kills them. It is not survived, and the fact that this patient is very obviously living excludes hemorrhage. It is only the subject of clinical demonstration at a much later period of life, because the hemorrhage that is survived is the result of degenerative changes practically confined to the degenerative period of life, and hemorrhage in the young is the result of disease of the larger arteries, such as permits their distention into an aneurism, the rupture of which is fatal.

Vascular obstruction it must therefore be with which we have to deal in this case. Veins or arteries may be obstructed; but the veins that are obstructed are on the surface of the brain, and we have found evidence that the lesion here is not at the surface, so we may put venous obstruction on one side. It must be arterial obstruction. This may be the result of two causes, which, widely differing in origin, have the same ultimate effect. They are embolism, closure by a plug brought from a distance, and thrombosis, closure by a clot formed *in situ*. Do not forget the distinction. The advice may seem superfluous, but it is curious how the idea of embolism dominates the minds of practitioners to the exclusion of thrombosis, and they call "embolism" whatever is not hemorrhage. But the distinction is quite as important as that between hemorrhage and obstruction.

An embolus must have a source. Almost invariably this source is a valve of the heart. To justify a diagnosis of embolism you must find a source,—that is, practically, you must find valvular disease of the heart, or, if the attack occurred some months ago, you must have a history of some malady, not long before the onset, known to cause endocarditis. Do not forget this also, because endocarditis, recently acute, may cause cerebral embolism, and yet may pass away so completely that at the end even of six months, not only may there be no murmur, but the walls of the heart may have so perfectly resumed their normal state that you cannot perceive any indication that there has been any cardiac disease. In this case no cardiac disease can be detected, and, further, no malady can be heard of that will cause endocarditis. Embolism may therefore be excluded, and we are left with thrombosis as the cause of the condition, thrombosis producing a focus of necrotic softening.

But we have not yet done with the process of diagnostic analysis. Thrombosis has two causes. It is a clotting of the blood, and it may be due only to a strong tendency of the blood to clot. This, however, is very rare. It occurs in the old and gouty; it occurs in the subjects of cancer; it occurs in states of profound general weakness, and it occurs especially soon after childbirth, when the vessels of the uterus have to be closed by clot, and the tendency to coagulation may be regarded as physiological,—physiological, but, unhappily, often also pathological, as the disastrous thrombosis in the pulmonary artery shows us too often. But here we have no evidence of such diathetic states, no history of general weakness, and, since the patient is a man, we may safely exclude the puerperal condition.

The second cause of thrombosis to which we are thus reduced is disease of the artery at the spot, disease which induces the formation of the clot. Such disease thickens the wall, narrows the lumen of the vessel, often almost to the point of closure. It also changes the inner surface, so that this acts like a foreign body, and on it a coagulum readily forms. The effect is produced in greatest degree on the branches that come off where the wall of the main artery is diseased. Beyond the narrowed part the vessel is wider, but the narrowing permits little blood to flow through, and hence the circulation in the wider part is very feeble, and where the narrowing ends we have the change in the wall. Hence chronic disease of the walls of the arteries leads to sudden occlusion by the formation of a clot, and gives rise to symptoms of sudden onset.

There are only two forms of arterial disease which act upon the arteries going to the central ganglia of the brain. A third cause, traumatic arteritis, only acts upon the arteries of the convex surface, and it is, moreover, extremely rare. You know, probably, what the two conditions are,—atheroma and syphilitic disease. Each has the same effect; in each there is a thickening of the wall, a narrowing of the vessel, and especially a narrowing of the branches which the main trunk gives off at the spot, for it is generally a main trunk which is the seat of the change. But atheroma is a senile lesion. It is a disease of the old, even more emphatically than is hemorrhage, for in extreme old age it becomes the more common lesion of the two. But this disease also the patient's age excludes.

By the mode of onset and the age of the patient, taken together, we thus arrive at the diagnosis that the cause of the symptoms must be an area of softening due to syphilitic arterial disease.

The time occupied by the onset, although the chief, is not the only guide. We may leave out of consideration those indications which apply only to patients in the later period of life, which we shall have another opportunity of considering. In the early period the distinction between the two common causes of hemiplegia, embolism, and thrombosis from syphilitic disease—for these two embrace certainly ninety-five per cent. of the cases of sudden onset in early adult life—is sometimes aided by the presence or absence of symptoms before the onset. Until the moment when an embolus is carried with the rushing blood into a vessel too small for it to traverse, the state of the intracranial structures is normal. In embolism, therefore, we have no premonitory symptoms. But it is not so in arterial disease. The narrowing of the arteries and the morbid process which is going on in their walls might be expected to cause symptoms constantly. It does so frequently, but not constantly, apparently because thrombosis occurs earlier, and with less extensive disease, in some cases than in others. One patient has headache, another has tingling or transient attacks of slight weakness in the limbs afterward paralyzed, the result of the interference with the blood-supply to the region from which it is afterwards altogether cut off. This patient had no premonitory symptoms in the limbs, but he had headache for a month, greater during the week immediately preceding the onset. So that this symptom yields us another indication pointing to the same conclusion as that to which we have been conducted by the study of the onset and the process of successive exclusion.

Loss of consciousness at the onset is chiefly important when the distinction has to be made between hemorrhage and softening. In the latter it is more often absent than present, and it was absent here. The only help it gives is that it is rather more often absent in thrombosis than in embolism, because the latter is more violently sudden, and it seems to give rise to a more definite shock to the brain. So this indication also affords confirmation of our opinion; although it is slight it corresponds in direction with the rest, a correspondence you should never fail to notice.

There are other occasional peculiarities of onset which afford a distinctly useful indication, and such is the onset of the symptoms in the patient before you. When the paralysis comes on in a series of distinct sudden attacks, each without loss of consciousness, each increasing the extent or degree of those which preceded it, we have a form of onset which is seldom, very seldom, met with except in thrombosis. The successive attacks mean the successive occlusion of branches of the diseased vessel. When hemorrhage has a deliberate onset it is gradually progressive from the slow forcing of blood into resistant tissues, or there are but two stages, of which the second is attended with profound coma and general paralysis, because it is the expression of rupture into the ventricles of the brain. Thus, in this patient the features of the onset, and its immediate antecedents, entirely confirm the opinion to which we arrived from the comparative study of the pathological possibilities.

Yet there remains another question of ultimate importance which may destroy our inference, leave it unchanged, or strengthen it. It is the question whether any evidence of the assumed cause of the morbid process can be ascertained. That, in most cases, is the form in which the question must be put. But if no cause of any one of the possible lesions can be traced, it is necessary to consider whether the cause of that which is otherwise indicated can or cannot be excluded. If it cannot, the indication may be followed. Hence, in the case before us, having found that syphilitic disease is the probable cause of the cerebral lesion, we ask, Has the patient had syphilis? To this the answer is an uncompromising negative. We can ascertain no history of any primary lesion or of any secondary manifestation of the disorder which we assume to be the cause of the morbid process which has brought him now under our observation. Some of you, and many of those who are not conversant with the progress of recent observation, would be inclined to accept the negation as a decisive disproof of our conclusion. They would be wrong. We have here a pertinent illustration of the importance of the absence of disproof where proof has been found wanting. Those who see many patients suffering from syphilitic affections of the nervous system, see among them some who, as this patient, can give no account of any chancre or of any secondary symptoms, and among these there are some in whom a minute and careful examination reveals some conclusive indication of syphilitic disease, it may be in the tongue, in the throat, on the face, or in the eye, but it shows how small is the practical importance of the ignorance which might seem to some to be conclusive. There are others in whom no such indication can be found, but who at some future time come under our notice with distinct later symptoms unmistakably syphilitic. Now, the existence of these two facts makes it impossible for us to accept the patient's belief that he has never had syphilis as destructive of our diagnosis. All that it does is to cause us to ask that second question, Can the patient have had syphilis? The disease is, for practical purposes and outside the ranks of the medical profession, acquired in only one way. If there has been no exposure to the risk of contagion, we may exclude the disease and we must be wrong in our diagnosis. If there has been such exposure to possible contagion we may be right, and a sufficiently extended observation compels us to regard the negative history as then without significance. It is so in this case. For five or six years the patient has been frequently exposed to the risk of contagion, and therefore we may regard our diagnosis as unaffected by the negation, and regard it as the expression of ignorance of the malady, not necessarily of its non-existence.

One or two features in the symptoms and in the lesion should not be passed by. The general symptoms of hemiplegia, as I said, we will defer, but almost every case presents some special features which we may not soon meet with again. Notice that the hand, although weak, presents no complete paralysis of any movement, but notice also that the movements are a little irregular. When he tries to touch his nose with his first finger, and eyes shut, he does not carry the tip of the finger straight to the tip of the nose. Such defective co-ordination is common with moderate recovery of power, and has not, as far as we at present know, any definite significance. It is not here associated with the symptom which some patients present, an ignorance of the position of the hand on passive movement. Next observe that sensation is unimpaired throughout the side. From this we infer that the posterior third of the hinder limb of the internal capsule has escaped, and therefore that the adjacent grey matter has escaped; and we are, therefore, prepared for another negative fact which you should always ascertain and note, that there is no hemianopia. There is no affection of sight, and there are no morbid changes in the eye. Syphilitic disease of the arteries and its results never cause optic neuritis. If there is optic neuritis you may feel sure that, in addition to the arterial disease, there is a syphilitic growth in some position in which it has not caused symptoms.

Lastly, look at the movements of the face, and remember that in all cases of facial paralysis it is necessary to observe the amount of interference with three different kinds of movement which take place in the face, especially in the lower part, which alone is involved in this case. The first is voluntary movement. When told to raise his upper lip and put his teeth together, you see that the movement is considerably less on the right side than on the left. The second is emotional movement, which is best manifested in the smile. But it does not do to tell the patient to smile. A patient who smiles to order produces only a voluntary and not a true emotional movement. To observe the latter you must produce the emotion. You may tickle a child, but you cannot well, without loss of dignity, adopt this method in the case of a grown-up person. But you may generally obtain what you need by asking the patient, as I now ask this patient, to "favor us with a *graceful* smile." You see that there is much less difference between the movement of the two sides than in the voluntary movement. Remember that this difference exists only when the disease is in the cerebral hemisphere, or at least above the middle of the pons. It is not seen in any case in which there is disease of the facial nerve, or its root fibres, or its nucleus. Lastly, the associated movement of the face, which occurs when the patient exerts force with some other part of the body, when he grasps with energy, for instance, may present a difference from the voluntary movement similar to that which is observed in the case of the smile. It may be far more equal on the two sides, and often no difference can be detected. We do not yet know the significance of these differences when there is disease in the hemisphere, but they show that the cause of the facial weakness is in this position, and they will probably ultimately prove to have a special meaning. The last point is the state of reflex action. The patient presents an increased knee-jerk and a foot-clonus on the right side; these, as you know, show that there is secondary degeneration in the pyramidal fibres. But the plantar reflex is almost absent on the right side while active on the left, and there is a distinct difference in the abdominal reflex, although it is less conspicuous, because, as is often the case in adults, this reflex action is not readily obtained. Such a contrast between the two forms of reflex action is common in hemiplegia, and is very remarkable. It is also mysterious, because we do not yet know to what the diminution of reflex action from the skin is due. It is present immediately after the occurrence of the cerebral lesion, while permanent increase in the myotatic irritability does not manifest itself until toward the end of the first week. We urgently need a series of careful observations on the position of the cerebral lesions in cases in which this diminution in the superficial action is present, and their position in the cases in which it is unchanged. It is not a recondite subject, and I would commend it to your notice.

Regarding the lesion, several points demand our notice. What is the change in the walls of the vessels which we have assumed to exist in this case? I show you under the microscope some sections of such diseased arteries, which have been prepared by Dr. Taylor. They show very beautifully the main features of the morbid change. You can see that there is a very extensive growth of cells and fibres in both the inner and outer coat of the vessel, the separation between the two being indicated by the wavy line of the elastic lamina. You may also see that the growth has almost closed the chief vessel, and that some branches which are divided in the section are completely occluded.

This disease of the arteries causes nodular swellings on the external surface, often considerable in their prominence. They are less translucent than the normal wall, but less opaque than the enlargement which occurs in atheroma. That is because, prone as syphilitic tumors are to undergo caseation, the tendency is slight in these arterial growths, and hence we have not the dense opacity which is due to fatty degeneration in the senile lesion. The change is especially common on the basilar artery and on the middle cerebrals, and in this case we can have little doubt that such disease in the left middle cerebral has occluded those branches which go to the central ganglia through the anterior perforated spot. It probably has not arrested the circulation through the main trunk, and so the cortex is intact and the patient has escaped the impairment of speech, which is the common result of the softening of the cortex, which occurs when the main trunk of this vessel is occluded.

But I would call your attention to one point, which is not, I think, remarked in any text-book. If the patient has been subjected to treatment, the disease in the wall of the artery is considerably changed. The swelling diminishes and may be scarcely recognizable, but the wall is still a little thicker and a little more opaque than normal. The aspect resembles that of slight atheroma far more closely than does the syphilitic disease in its recent and unaltered state. You must therefore be prepared for this appearance in the case of many patients who have been under the care of those who recognized the cause of the symptoms. Often you find such old disease, altered in the manner I have described, in one artery, and in another artery recent change of characteristic aspect. Not long since I met with a very instructive example of this, most instructive, also, in its clinical manifestations. The patient was a doctor, thirty-five years of age, of reticent habits, who had suffered from a sudden brief attack of loss of speech. Six months before, he was attacked with sudden symptoms of an extremely grave character. Rising in the morning well, although there was some reason to believe that he had suffered from headache for a few weeks, he was suddenly attacked with difficulty of articulation and with mental stupor, deepening in a few hours almost to coma, and attended by paralysis of the right arm, right leg, and lower part of the face, complete paralysis of the muscles of the left eyeball, and almost complete of those of the right, with immobility of the pupils, the right being small and the left dilated. So he continued for two or three days, and then, with rising temperature, he died. The symptoms pointed conclusively to a lesion of the oculo-motor nuclei beneath the corpora quadrigemina, greater on the left side and extending through to the motor tract, a lesion of sudden onset, and therefore vascular, and, from the reasons we have considered in our review of the case before us, certainly due to vascular occlusion, to arterial disease, and almost certainly to syphilitic disease of the basilar artery and posterior cerebrals. The attack six months before might reasonably be ascribed to similar disease in the middle cerebral of the left side, disease which, on the assumption that he was conscious of previous syphilis, would naturally lead him to take iodide of potassium, the influence of which would be to prevent, for the time, further effects of the disease. He could easily have obtained it, and would have been unlikely to communicate his apprehensions or self-treatment to any person. The omission of the drug, or possibly, even, as we shall presently see, its too long continuance, would have permitted a fresh development of arterial disease in the basilar and posterior cerebrals, with resulting and fatal thrombosis within them. Such was the opinion expressed on the first examination of the patient, the day after the onset, and the subsequent condition afforded me no grounds for changing the diagnosis. The post-mortem examination showed its correctness in a degree which would be remarkable, were it not that the elements of the diagnosis were simple, and only required trust in the process of reasoning to induce confidence in the conclusion. There was thrombosis of the front of the basilar, extending into the posterior cerebrals, further into the left, and on this there was a characteristic nodule of syphilitic disease, while slighter thickening, narrowing the cavity of the vessels, was seen on the other posterior cerebral and on the front of the basilar. In the middle cerebral, within the fissure of Sylvius, was an area of altered wall of greater opacity and thickness, but without much prominence, precisely the aspect I have mentioned to you as left by syphilitic disease which has been removed by treatment, and justifying the opinion that I have mentioned.

What is the relation of the vascular lesion to the constitutional affection? Although not yet demonstrated, it is impossible to doubt that syphilis, like every disease that is transmitted by inoculation, that has a period of incubation and is first manifested by exanthematous and allied symptoms, depends on a specific organism which multiplies in the system and produces germs. As far as we can read the meaning of the course of this malady, it would seem that the various morbid processes which it causes are the result of the presence of developed organisms, abundant in the blood in the exanthematous stage, and leaving germs which rest in some tissue until a time comes when they develop and give rise to a process of tissue-growth such as that which we have been considering in the walls of the arteries.

Accumulated and accumulating experience leads us to the conclusion that the remedies employed for this disease —mercury in the first rank and iodide of potassium in the second—destroy the developed and developing organisms, and remove, or promote the removal, of the tissue-growth which the organisms have caused. Hence the manifestations of the disease in the early florid stage, or the later occasional effects, may be with certainty swept away. But the agents seem to have no influence on the germs which are deposited in the system and are not yet in the process of development. You are doubtless familiar with the fact that the germs of bacterial organisms have far greater power of resistance than the organisms themselves; that they resist, for instance, temperatures which are fatal to the latter. Hence the removal of the symptoms at any stage of the disorder does not free the patient from the liability to the occurrence of future symptoms, the result of the development of the germs that have been untouched by the therapeutical agent that has been employed. The most thorough treatment at any stage of the disease does not prevent its future manifestations, and the freedom from these is as frequent in those who are not so treated as in those who are. It depends on the disease, on the amount of organized material in the system, and on no other element. It is in this sense that I have expressed my conviction that syphilis as a disease is not curable. I have been taken severely to task for the assertion, but it has received no dissent from those who know most of the malady, and whose opinions deserve chief attention. I believe that the statement is the assertion of an incontrovertible truth.

All that we have hitherto considered is but the path to the problems that are of practical moment,—those that concern the interest of the patient,—and, therefore, to us as practitioners are of paramount importance. These are the prognosis and the treatment, the attempt to forecast the future, and the attempt to remove or lessen the symptoms and to prevent their return.

Strive, gentlemen, when you consider the prognosis of such affections, to form a mental picture of the morbid process. Note its elements, their mutual relation, and their relation to the symptoms. You may perhaps be surprised to perceive, when you do this, that in all cases of true syphilitic lesions of the nerve-centres the symptoms do not directly depend, in every case, upon the syphilitic process itself. This is conspicuous in the case of the disease with which we have to deal to-day. The symptoms depend on the necrotic softening of the brain, which is precisely the same as that which would be caused by any other form of arterial closure, by embolism, or by thrombosis due to any other mechanism. It is the result of the arrest of the blood-supply, and the clot which finally stops this is also common and not specific. It is only when we reach the cause of the narrowing of the artery that we reach the actual syphilitic element in the process. But you cannot fail to perceive, further, from these facts, that the course of the symptoms must be altogether independent of the course of the arterial disease which led to the obstruction. Remove the disease in the artery; it still remains impervious, and ultimately becomes a mere fibrous thread; nothing can restore its cavity and the destroyed brain-tissue also nothing can renew. Such a patient will improve, partly by the recovery of the least damaged structures, partly and chiefly by the compensation which happily the other hemisphere is able to effect in the case of all movements which are related to both sides of the brain, those in the upper part of the arm and face, and the chief movements of the leg. But the future course of the paralysis, in every case of the kind, will be the same as if the obstruction were the result of embolism, and the fact must be frankly recognized. Do not, as is so often done, assure a patient with hemiplegia of a year's duration that he will recover, because his malady was the result of a process which you know can be removed, and which most likely has already been removed.

Is treatment therefore useless? By no means. Its importance is great, and the greater the sooner after the onset you can apply it, and it is greatest of all before the onset, when the shadow of the coming disaster is thrown by premonitory symptoms, and when the substance may very often be averted by the prompt recognition and energetic treatment of the morbid process. I am certain that in cases which have come under my notice this result has been achieved, and palsy, perhaps life-long, and sometimes even death, have been averted. In this patient we have an illustration of the opportunity which many others present: when he first felt the disability in the arm, the merest suspicion of its possible cause would have justified treatment which would certainly have averted the sequel. After the closure has done its work, and the softening has produced its symptoms, treatment can do nothing for that which is, but it may still do much for that which may be. We cannot tell how many other arteries are diseased, how many other branches are in imminent danger of being closed, and it is essential, therefore, at once to give the patient iodide of potassium, seven or ten grains three times a day, and even fifteen grains for the first few days, with, if you like, mercurial inunctions. I believe that this dose of iodide is adequate, and will do all that can be done in the course of four or six weeks. It is better not to give larger doses, because coagulation of the blood plays a part in the mechanism of destruction, and large doses of iodide, as the treatment of aneurism has taught us, increase the tendency of the blood to clot. Of even more importance is it not to continue the iodide beyond six weeks or, at most, two months. In that time, as visible processes show us, all the specific element in the process is removed. If recovery is not complete, it is because the simple elements may take yet a longer time to pass away, or may be of such deeper character that they must persist. If you go on with iodide for four or six months, you may find the process that was at first arrested regains activity, and the very same syphilitic process may actually kill the patient, in spite of the continuance of the treatment which at first arrested it. Apparently the organisms become acclimatized to the presence of the agent, are able to resist it, and thrive in spite of it. Analogous phenomena are known, as I have hinted in the case of similar organisms and the effect of high temperatures. But while you should not continue the treatment, you should always resume it. After three or four months' cessation it is as effective as at first.

I consider that every patient who has had syphilis should for at least eight years from the primary disease, or five from the last manifestation of it, take a course of iodide for three weeks twice a year. In this way developing organisms which have not yet caused tissue-changes sufficient to produce symptoms may be destroyed, and in many cases it is certain that this measure would save the patient from grave disease. We cannot see the result, but in this it is but on the level of all the other forms of preventive medicine, the branch of our professional work to which we may look up with the highest pride, and go on with, confident in its unseen achievements, despite the fact that the recipients of its greatest blessings are all-unconscious of them, as they are always of the health they enjoy until they lose it, and of the air they breathe until the last breath may bring to their mind the flash of revelation, all too late.

LECTURE V.

BULBAR PARALYSIS.

Gentlemen:—I wish to-day to take the opportunity that this case affords me of showing you the symptoms of the malady generally known as "bulbar paralysis." It was formerly called by a descriptive designation, given to it by Duchenne, " labio-glosso-laryngeal paralysis." The symptoms in this case are so complete in degree that they may impress upon your memory the chief features of this malady. You are not likely to forget them after you have observed a well-marked case.

Let us first briefly note what they are. (1) You will observe, when I tell the patient to move his lips and mouth, that he has but little power of moving the lower part of the face; although he can put in perfect action the muscles of the forehead and eyelids. He can only slightly raise the upper lip; he cannot narrow the orifice of the mouth; although he can just approximate the lips, it is imperfectly; a slight chink remains between them, and they cannot be kept together. (2) His tongue is almost completely paralysed; when he tries to put it out, you see that he cannot get the tip beyond his teeth. Moreover, if you look at the surface of the tongue as it lies in his mouth, you will see that it is uneven; irregular depressions upon it indicate the atrophy of the muscular tissue of which it chiefly consists. (3) His palate can be raised, but if he tries to swallow, liquids come back through his nose, evidence that the palate does not shut off, as it should, the cavity of the nose, on account of feebleness of the palatine muscles. But semi-solid food is swallowed well; it does not regurgitate into the nose, and when it has once reached the pharynx it is passed down into the œsophagus without difficulty. This is proof that the pharyngeal muscles are not paralyzed. (4) It is very different with the muscles of the larynx. When he tries to phonate he only succeeds in making one uniform vowel sound, not altered by any changes of intonation. When he tries to cough, you can readily perceive, by the sound, that he does not close his glottis; there is a rush of air through the larynx, but there is no true cough, for which, you know, the glottis needs to be first closed and then suddenly opened. The paralysis which is thus indicated can be seen by the laryngoscope. Dr. Semon has been good enough to examine his larynx, and finds that the left vocal cord is near the middle line, and cannot be moved away from it, but the right is habitually some distance from the middle line, so as to leave a space of about 4 mm. between the two. When he tries to bring the cords together, the right moves toward the left, but does not reach it. In the movement, both cords present a peculiar tremor. The left cord thus presents abductor paralysis, while in the right adduction is most affected. It is fortunate for him that there is not loss of abduction on both sides. If there were, he could probably cough and utter vocal sounds much better, but his life would be in constant danger, because, as you doubtless know, when the cords cannot be separated, the slightest catarrhal swelling suffices to close up the narrow space that remains, and, unless an opening is made into the larynx below the glottis, breathing becomes impossible, and the inevitable result ensues.

So this man presents paralysis of his larynx, palate, tongue, and lips, together with the lower part of the face. The malady has come on gradually, in the course of the last two years, and the first part to suffer was that which is now most affected, the larynx. He first found that he could not sing; then that he could not speak well, from weakening of the tongue; gradually the other symptoms were added. Thus, it is an example of slow gradual bulbar paralysis, a malady which we know depends upon a slow degeneration of the structures of the medulla oblongata concerned in the movements for articulation. Let us glance briefly at the parts that are affected. In doing so I would ask you to notice how remarkably the affection corresponds to their functional use. Those are paralysed which are concerned in articulate speech. Some of these are also concerned in deglutition, and hence swallowing is to some extent affected, but the fact that the pharnyx is not paralysed is proof of the manner in which the functional limitation of the affection has spared the special act of swallowing. We shall presently see the reason for paying such close attention to the relation of the symptoms to function. It is a point of great importance because it is of great significance, and you should always ascertain how far it is to be traced. It is one of the chief features to be attended to in each case that comes under your notice.

This bulbar paralysis is generally a disease of the nuclei of the nerves which convey motor power to the muscles affected, and when such, it is a disease of the medulla oblongata analogous to the degeneration of the motor nerve cells of the spinal cord which gives rise to the symptoms of "progressive muscular atrophy." In the medulla, you will remember, the regular arrangement of nerve cells, root-fibres, and nerve-roots, which obtains in the spinal cord, gives place to a very irregular arrangement, in which the fibres which subserve a certain function are, for the most part, gathered together in a single nerve, and rise from a defined group of cells, the nucleus of the nerve. The irregularity is indeed far greater in the case of the sensory than in that of the motor fibres, for almost all the sensory fibres leave the medulla as a single nerve, the fifth, which represents the posterior roots of all the motor cranial nerves. But the irregularity is chiefly in the way it leaves the brain, for within the pons the fibres of the nerve at once separate; some ascend and some descend, and all seem to end in sensory nerve cells at about the level of the motor cells to which they are related, as do the sensory fibres of the spinal nerves. But the malady we are now considering is one of motion, not of sensation, and so we may dismiss the sensory fibres of the fifth nerve from further consideration, and attend only to the arrangement of the motor nerves and nuclei, whose function is so conspicuously impaired.

As we pass up the cervical region of the cord, before we reach the medulla, we have, as it were, an intimation of the coming irregularity. The nerve-fibres for the muscles of the neck in part arise, and leave the cord, as do the motor fibres lower down. But some of them, arising from the same anterior grey matter, pass outwards instead of forwards, and leave the side of the cord, instead of the front, to form by their junction a nerve which ascends to the medulla and passes within the foramen magnum as if it were trying to attain the dignity of a cranial nerve. It does actually join one of the bulbar nerves for a short distance, and so this is called the "spinal accessory," but its fibres have to separate from the bulbar nerve, and descend to the neck, where, as you will remember, it supplies the sternomastoid and the trapezius, muscles that serve to rotate the head. I mention this nerve thus specially, because the bulbar nerve, which its fibres join for a short distance, particularly concerns us. The grey matter from which it comes would be continuous with that from the neck muscles, had not its relation to the other parts been deranged by the fibres of the anterior pyramids of the medulla, which here cross the middle line and cut up the anterior cornua. The cells for the spinal accessory in the medulla are the lowest of the bulbar groups. Just as the spinal part of this nerve goes to the external muscles of the neck, the bulbar part goes chiefly to the internal muscles of the neck, the muscles of the larynx. But you perceive that there is a great difference between the two parts in source, in destination, and in function, and this difference is reproduced in disease. In the patient before you the outer spinal part of the nerve is not paralysed; the inner, bulbar part, as we have seen, is almost completely paralysed; the cells of its nucleus must be extensively diseased. If we pass upwards still higher, in observing the parts affected, we come to the palate; this is supplied by the same nerve, the spinal accessory, but probably by fibres that arise from its highest cells. The supply of the palate by this nerve has only lately been proved, but it has been proved beyond the possibility of doubt, and the fact is of much interest in connection with such a case as this.

Next, after the affection of the palate, we pass to that of the tongue, which is so severe. The hypoglossal nucleus lies parallel to that of the inner part of the spinal accessory, at first below it, afterwards to the inner side of it. Not only are the nerve cells for these three parts near together, but so also are their nerve fibres, as they leave the medulla, and the tongue, palate, and vocal cord are often paralysed together by a morbid process at the surface, damaging the nerve roots.

How closely structural relations reproduce those of function, how, indeed, they determine the latter, is again illustrated by the last part of the combined paralysis which we have to consider, that of the lower part of the face, and especially the lips. The chief nucleus of the facial nerve is only a little above the upper extremity of the hypoglossal. The nerve leaves the pons above the level of its chief nucleus, to which its fibres pass downwards by a somewhat circuitous course. But the relations of the fibres of the nerve which subserve its particular functional relations are still, in the main, unknown to us. The muscles supplied by it have a twofold function (1) in connection with articulation, in which the lower functional muscles are chiefly concerned, and, in particular, those of the lips. These are the muscles which are so much affected in the case before you. (2) The muscle which closes the eyelids is concerned, in its function, with the muscles of the eyes, and this is also true to some extent of the muscles of the forehead, for when we look up the frontales raise the eyebrows, and when the eye has to be protected by a strong contraction of the orbicularies, the corrugator assists. (3) All the muscles of the face are concerned in the expression of emotion, the upper as well as the lower muscles of the face. It is only the first of these three functional relations which concerns us now.

It is certain that there must be a close connection between the fibres or cells which have to do with the muscles about the mouth, and with those for the tongue, but the anatomical connection has not yet been traced. That there is such an anatomical connection, we cannot doubt, nor can we doubt that it is especially diseased in such a case as this. Apart from the evidence of the relation afforded by disease, I may call your attention to a curious proof that is to be obtained, even under normal conditions, of the closeness of the relation between the tongue and the lips, and also between the fibres or cells of the facial and hypoglossal nerves which innervate the parts. Try to narrow your tongue; you will find you cannot do so without, at the same time, narrowing the opening of the mouth. You cannot contract the transverse fibres of the tongue without also contracting the transverse fibres which run in the lips. If you try to do either of these you will find that you do the other also. The experiment succeeds best, or at any rate we are most conscious of it, if the tip of the tongue is first placed between the lips. Try it now, and your sensations will speedily convince you of the truth of the statement.

Thus the symptoms are dependent on a paralysis of the muscles supplied by part of the facial, part of the spinal accessory, and by the hypoglossal nerves.

The paralysis corresponds to function in its distribution, and has been gradual in development. These two features are always evidence of the degenerative nature of the process on which the paralysis depends.

But it is important to note that there are two forms of this chronic bulbar paralysis. I have told you that these nerves and their nuclei, although situated in the medulla, correspond to the motor nerves and motor grey matter of the spinal cord. The same correspondence is to be traced in the chronic palsies which affect the two parts. In the limbs we may have the same two forms of paralysis. We may have the slow paralysis confined to the wasting muscles and dependent on the degeneration of the grey matter of the spinal cord from which the motor fibres proceed. We may also have a slow paralysis without wasting, but with excess of the muscle-reflex action, and this, as you probably know, depends on a slow degeneration of the fibres which conduct the voluntary impulse from the brain to the spinal grey matter. In some cases of chronic bulbar paralysis we have conspicuous wasting of the parts paralysed, and then we always find degeneration of the cells of the nuclei from which the fibres proceed, especially conspicuous in the cells of the hypoglossal nuclei. This form is often associated with the corresponding muscular atrophy in the parts supplied by the spinal nerves, especially in the arms. In such a case, however, we have no wasting, but a slow paralysis similar to that which I have described as met with in the limbs. Although the precise lesion in this case has not yet been made out there can be little doubt that it depends on a degeneration of the fibres which convey the voluntary impressions from the brain to the nuclei. These pass with the fibres for the limbs until a short distance above the nuclei, and then they pass across the middle line to reach the nerve cells for which they are destined. The fibres for the limbs cross the middle line chiefly at the "decussation of the pyramids."

The only explanation of such cases that we can give is that, in them, these upper fibres suffer just as do those in the spinal cord in primary lateral sclerosis. In the spinal cord the degeneration probably begins at the lower extremity of the fibres and extends upward, precisely as it does in the peripheral nerves in some forms of multiple neuritis, and we may reasonably assume that the affection of the similar fibres for the nerve-nuclei of the medulla has a corresponding course. If so, we are able to understand the strict correspondence with function which may be often discerned in such cases, as well as in those in which the nuclei themselves suffer and the muscles waste. Even in the case of the degeneration of the peripheral nerves, the correspondence with function is often remarkably close. But in both forms of bulbar paralysis the degree of correspondence with function varies. In the patient before you it is strict; in the case from which I am about to show you a section, not only the muscles in articulation, but also those in deglutition, were affected; the pharynx was paralysed as well as the larynx. We are not able, at present, to attach any special significance to this difference; it is possible that some day we may find in it an important indication of the cause of the disease.

These then are the two forms of chronic bulbar paralysis. Because the nuclei that are affected are those for the lower cranial nerves, it is sometimes called "inferior nuclear paralysis;" the group of nuclei in the upper part of the mid-brain, those which innervate the eye muscles, are sometimes affected in a similar manner in what is called "superior nuclear paralysis." It is indeed "superior" only as regards the chief nucleus affected; in this we have an interesting instance of the way in which function, rather than locality, determines the grouping of degeneration. The nucleus of the sixth nerve, you will remember, is quite low in the pons; so low, indeed, that the facial nerve, which is affected in the inferior palsy, actually winds round it, and yet this nucleus is involved with that of the third nerve in the superior nuclear palsy giving rise to the affection which has also been termed "ophthalmoplegia externa." In other rare cases both sets of nuclei suffer, generally, however, incompletely. Some of you may remember a woman who has attended here, who has a considerable degree of superior nuclear paralysis and a slight degree of inferior paralysis. I shall have to allude to her case again, because it is remarkable in being a sequel to diphtheria.

The two forms of chronic disease which I have described do not include all the cases of bulbar paralysis. There is also an acute form, or rather, I should say, a sudden form, which depends upon the closure of some of the arteries which supply this part with blood. The closure is the result of disease of the main trunk from which the small arteries spring; thickening of the wall necessarily entails the narrowing of the orifice of the branches that arise there. Most of the blood comes from branches which pass near the middle line, either from the vertebral or from the basilar artery. The resulting paralysis is generally much more irregular than in the degenerative chronic variety, as we might expect it to be, since the effect of vascular disease is always to cause a more or less random lesion. When one vertebral is diseased, and the branch which is closed comes from it, the affection is chiefly one-sided. Now and then, however, even when thus produced, it is bilateral, and the explanation of this, at first puzzling, fact is to be found in the frequent disparity in size of the two vertebrals. This disparity may be such that the nerve nuclei near the middle line, which are those especially diseased, are supplied with blood on both sides by the branches which come from the one large vertebral artery. Then, of necessity, disease of that artery and closure of the branch cause symptoms which are bilateral and symmetrical. I will show you in a moment an example of this instance of bulbar paralysis.

There is yet a fourth variety to be considered, which is, however, so rare that it may be dismissed with a mere mention. Very, very rarely these nuclei seem to be the seat of a true inflammation, and we then have a bulbar paralysis which is not sudden but acute in onset, as the effects of all inflammations are. Such a lesion is extremely rare, but it is a little less so in the case of the superior group of nuclei. It is important, however, to mention it, because it has been known to supervene on a chronic degeneration, and to be an immediate cause of grave danger. Indeed, if the list is to be made complete it is necessary to mention that mysterious cases have been met with in which the characteristic symptoms of bulbar paralysis have existed and have caused death, although no morbid state could be discovered either in the medulla or in the nerves. We can only explain such a fact by supposing that there are grave alterations in the nutrition of the nerve elements, sufficient to abolish their function, which are at present not within the range of our means of observation. We know that this is the case in chorea and also in paralysis agitans; although in these affections the changes do not abolish function, it is quite conceivable that, in the almost inconceivable complexity of the molecular nutrition of the nerve cells, alterations may occur which do arrest function, and which yet entirely elude our present means of detecting them. But it is also possible that in some of these cases there may be a degeneration of the fibres which conduct the motor impulse from the brain to the nuclei, and are, in a sense, homologous with the fibres that pass to the grey matter of the spinal cord in the pyramidal tracts. We have reason to believe that, in the latter, the process of degeneration only begins at their lower extremities, as it does at the lower extremities of the peripheral nerves in most forms of multiple neuritis. In such cases degeneration, chronic or acute, may give rise to symptoms of considerable severity, although confined to the terminations of the fibres, and thus the lesion may escape detection.

Let us now consider in greater detail the facts of the malady which is before us, the chronic degenerative bulbar paralysis. We know very little of its cause. It is most frequent in the old, but it occasionally attacks the middle-aged, as in this instance, and now and then, for some reason or other, it occurs in those who are still children. Some day we may know the reason of this. At present all we can do is to recognise what may be termed the outlines of the general causation of such affections. These diseases are degenerative. A short time ago this statement would have taught us nothing. But we have reason, of late, to recognise in most degenerations the evidence of the previous action of a toxic influence. Note this important law, that the affection of the nerve elements related to function, when acute, indicates the action of a toxic influence; when chronic, a degenerative process. It is a law of important practical application, but it has also an important pathological significance. The two things are not entirely separated. The toxic influence which has no action on the nerve elements may yet leave behind it some effect, possibly some chemical product, which fixes itself on the nerve elements, in consequence of which these nerve elements, at some future time, undergo degeneration. This has not been proved to be true in the case of bulbar paralysis, but it is very probable that it will be proved, at any rate in the case of the forms which occur comparatively early in life, and cannot be attributed to merely senile changes.

We see this connection between an early toxic blood state, and a late degenerative process, conspicuously in the case of syphilis and tabes. The same relation is to be traced in the case of the superior nuclear paralysis, the degeneration of the nuclei for the motor nerves for the eyeball of which I spoke just now. This may be associated with tabes, and it may also occur as a late sequel of syphilis apart from tabes, a fact of considerable significance. But we cannot trace its relation in the case of the inferior nuclear palsy, bulbar paralysis, which especially occupies our attention to-day. But some of you may have seen a patient who attends here occasionally, and who suffers from the double affection, from considerable paralysis of the ocular muscles, and from slight bulbar paralysis, in whom these symptoms have slowly developed as sequel to diphtheria.

We must, however, separate the rare cases that are met with before or soon after the completion of development. Allied facts suggest that the cause of these is different. They may be analogous to the mysterious "hereditary ataxy," "Friedreich's disease," and depend on some congenital defect of vital endurance in the structures concerned. But to pursue this would take us too far to-day. The possible relation to toxic influences is a subject of great importance, because at present we have little power of checking the developed disease, but if we can discover such a cause, the discovery may bring with it some means of prevention or arrest, of which at present we have no indication. One other fact deserves mention as bearing upon this matter, and that is, that I have once known characteristic bulbar paralysis to follow ordinary lead poisoning. The patient was in the late period of life, and so may have been disposed by mere senility to a degenerative process, but this chemical poison is known to be capable of giving rise to a similar process in the motor grey matter of the spinal cord, and it must be regarded as a very probable cause of the bulbar paralysis in the case in which the sequence was observed.

The general diagnosis of bulbar paralysis resolves itself into certain problems, or rather certain general laws. First, it is practically only the sudden forms in which a distinction has to be discerned from "pseudo-bulbar paralysis"—a condition which I have not been able to describe to-day. It is the effect of a lesion in each cerebral hemisphere, so situated that the second prevents the compensation for the first which is commonly effected. The only chronic process which may simulate the bulbar degeneration is the slow compression of the medulla, or its slow damage by growths inside or outside of it, and the symptoms then produced are always irregular, and combined with other, generally more obtrusive, disturbances, which sufficiently indicate the actual nature of the case. But the diagnostic difficulties presented by the sudden form are numerous. The double cerebral lesion generally declares itself by the closeness of two pronounced attacks of hemiplegia, first on one side and then on the other, the first of which has the common features of cerebral hemiplegia in the affection of the tongue and face on the same side as the limbs. It is not uncommon for sudden bulbar paralysis to come on in two or three stages by the successive closure of different arterial branches, but in these the paralysis of the tongue and lower part of the face is much more irregular than it is in the case of the cerebral lesions. The distribution of the symptoms in these cases is indeed so irregular as to make it almost impossible to give any general description of them, and this very irregularity constitutes their most important distinguishing feature. In the cases in which the symptoms are not irregular, their symmetry is generally conspicuous from the first, for the reason that I have explained,—because the nuclei on both sides receive their blood from the branches of a vessel on one side. But diagnosis of the precise nature of the sudden vascular lesion which causes the symptoms is often a difficult matter. When there is syphilitic disease of the whole of the vessel from which the nutritive branches proceed, the recognition of the fact is indeed seldom difficult. It is determined by the same evidence which enables us to recognise such disease in other arteries, and therefore need not detain us now. Embolism we may also dismiss as rare in this region, and associated with a distinct source of the obstructing plug. But the point of chief difficulty, and also of chief importance, is the determination of the question whether a senile lesion is hemorrhage, or softening from atheromatous thrombosis. The latter process is the common one here, and it may be safely assumed when the symptoms come on in several distinct attacks, as was the case in the patient before us. But the difficulty is rendered greater by the fact that a good many cases which have been described as hemorrhage have really been cases of arterial occlusion with secondary extravasations in the affected region. The distinction of such a lesion, in its early stage, from a primary hemorrhage, is not easy when the affected areas are small. A hemorrhage which is sufficiently extensive to cause considerable paralysis would certainly give rise to severe initial symptoms, and if these are absent it is justifiable to assume atheromatous thrombosis rather than the rupture of an artery.

The prognosis and course of the sudden and the chronic forms of necessity present a considerable difference. The chronic form, at whatever age it occurs, has a progressive tendency, and as long as the symptoms continue to increase serious anxiety cannot but be felt regarding the issue. Many cases of this description end fatally in one or two years from the onset. But in this disease, as in analogous progressive muscular atrophy, there is a marked tendency for the progress of the lesion to undergo arrest after it has reached a considerable degree of intensity, and has produced so much damage that the state in which the patient is left is pitiable. It seems to be the case in the patient before you, in whom the malady seems no longer to be increasing. The prognosis is certainly worse in a case in which the disease of the medulla supervenes on degeneration in the spinal cord, and bulbar paralysis is thus added to pre-existing muscular atrophy. It is worse, moreover, in the cases in which the symptoms are extensive and involve the parts concerned in swallowing as well as in articulation, the pharynx as well as the tongue and lips, than in the cases in which, as in this patient, only the structures for articulation are each affected. But of necessity the prognosis may have to be modified by the state of the larynx, and by the tendency there may be for particles of food to pass into the glottis and trachea. The chronic inflammation of the lungs, which is thus produced, is one of the most common causes of death, and when this source of danger exists the prospect of life being long preserved is always considerably lessened. From the larynx the chief risk is that which arises from the occasional paralysis of the abductors of the vocal cords on each side. In this patient, fortunately, such paralysis exists on one side only, and so all its characteristic indications are absent, with them the danger which bilateral paralysis always involves. When both cords are close together—so close together that the inrush of air in breathing gives rise to inspiratory stridor—the danger of suffocation from slight catarrhal swelling of the cords is, as I have said, extreme, and adds much to the gravity of the case.

In the acute form, if the onset is survived, some degree of improvement may be confidently anticipated, and it may be expected to go on for a considerable time. The chief danger arises from the existence of conditions favoring other attacks; it varies according to the probable amount of disease in this and other cerebral arteries, and is also influenced by the extent to which other influences have co-operated in producing the obstruction. It is governed by the same rules as similar softening in other parts of the brain, and I need not advert to it here. The danger of later extension is never entirely absent in atheroma, and even in the case of syphilitic disease it may continue for some time. The cicatricial process in the wall of the diseased artery, by which the disease is removed under the influence of treatment, may involve the formation of fibrous tissue in the wall, which goes on contracting, and this may entail the closure of some other small branch, months after the occurrence of the primary lesion, and after the disease of the wall has yielded to the influence of the drugs which are administered. Thus in all cases of sudden bulbar paralysis, the prognosis should be tinged with caution, and sometimes, even after the onset is survived, with distinct concern.

The rare cases of acute inflammation, in which the symptoms develop in the course of a few weeks, must at present be regarded as very serious. Ignorant of its causes, we are without the means of arresting the morbid process, which, in most of the few cases hitherto observed, has gone on to a fatal termination. But the probability that the process is due to a toxic influence is very great, and here also we may hope that the future will soon bring that which the present does not afford, and that we shall not have long to wait before we obtain the means of arresting the disease, and of doing so before the change in the nerve elements has proceeded to a degree that precludes their recovery.

You will have gathered from what I have said that the treatment of bulbar paralysis is one of the gloomy regions of the therapeutics of the nervous system. It is a subject on which there is, unfortunately, only too little to say.

That of the sudden form does not differ in any material point from the treatment of similar vascular lesions in other parts of the brain. When the sudden paralysis is followed by loss of faradic irritability, with preservation of voltaic irritability, that is, the reaction of degeneration, it is desirable for a time to keep up the irritability of the muscular fibres by the application of voltaism in any part to which electricity can be applied, in the hope that the nuclei from which the degenerated nerves proceed may be only damaged and not destroyed, and that recovery, with regeneration of the nerves, may ultimately occur. In such a case we follow, therefore, the same rule which would guide us in the treatment of an acute lesion of the motor grey matter of the spinal cord, and endeavour to keep the muscles in as good a condition as possible, in case the motor impulses should ultimately be again able to reach them and to act upon them.

In the chronic form, which occurs in the old, treatment directed to the morbid process seems to be entirely useless. The disease is a local senile change, and as such is beyond the reach of any influence we can bring to bear upon it or which the future is likely to afford us. The facts at present available regarding the chronic forms, which occur at earlier periods of life, suggest the same correspondence with analogous affections of the spinal cord and a similar practical application of the correspondence. In the form in which the muscles waste, the hypodermic injection of strychnine, of unquestionable service in the spinal affection, should always have a thorough trial. There is no reason why it should not be effective in the atrophic bulbar paralysis as well as in "progressive" muscular atrophy. Its use is not contra-indicated by previous administration by the mouth, and should not be delayed, because, if arrest can be obtained before the malady is advanced, even if improvement is not secured, the condition of the patient is far better than if the disease had obliterated most of his power of articulation, and some of his ability to swallow. The danger of pulmonary consequences, which result from the latter, must also be kept in mind in the treatment of every form of bulbar palsy, and every expedient, such as slow, deliberate eating, and food in a semi-solid form, which may lessen the chance of the entrance of particles into the trachea, and thence into the bronchi, should be insisted on. By these means, death may be averted or postponed more frequently than we can perceive. I have unfortunately not found anything effective in diminishing the flow of saliva, which is so great a trouble to many patients.

In general management, do not forget that the sufferers are often much distressed by the inability to make themselves understood; endeavor to lessen this as far as you can, by persuading them to express themselves in writing rather than by efforts to speak, which only end in failure, and involve a strain on the affected structures which cannot be other than injurious.

LECTURE VI.

FACIAL PARALYSIS.

Gentlemen:—Complete paralysis of the right side of the face, in a child of seven, who presents no other symptoms, —that is the problem before us. You have heard the questions asked, and the answers given by the child's mother, and that we found no evidence of a cause; you heard the mother give a negative answer to every inquiry, —there had been no blow, no exposure to cold, no discharge from the ear. Those are the three chief causal facts to be sought in every case. Even pain in the region of the nerve was denied. But one other question was asked that should never be omitted in such a case: Has the child suffered from earache? The mother then remembered the fact that had been forgotten, remembered it with surprise, so little had she thought it connected with the affection of the face.

You see the symptoms which the child now presents, and you have heard the story of the onset.* You see that the face is symmetrical at rest, but in movement all symmetry is lost in the distortion produced by the limitation to one side of all the familiar movements that the will can cause, or by which emotion is expressed. There is no movement, voluntary or emotional, in any part of the face.

Paralysis of all the muscles supplied by the facial nerve, on one side only, and without other symptoms, always means disease of the nerve trunk. Practically, moreover,

Clinical Journal, February 14, 1891.
* From a report by Mr. H. Caiger.

if it occurs without obvious disease or injury near the nerve after it emerges, it means disease of the nerve during its passage through the bone. These two facts should be fixed in your mind, and the reasons for the conclusion should also be clearly understood. Never learn a diagnostic rule, indeed, never accept any general assertion, without also endeavouring to ascertain on what the rule depends, or the assertion rests. <u>Unreasoned conclusions are the bane of students</u>. Without the facts to make them cohere with our previous knowledge, such conclusions will soon slip from the memory, or if they do happen to be retained, it is only by dint of pressure which alters their form, or drives them into some place to which they do not properly belong. They are remembered wrongly, and do you more harm than good. But fix the assertions by their evidence, the rules by their reasons, and not only do they remain, but they take root, and they become part of your real knowledge and a source of increasing power.

You are no doubt familiar with the distinction between the two chief forms of paralysis of facial muscles,—the general palsy of all parts which is often, as here, absolute, and the form in which only the lower part of the face is involved, which is never absolute. The first, as you know, means disease of the nerve, or of the nucleus from which the nerve comes, while the other (so common as part of hemiplegia) generally means disease of the cerebral hemisphere, and always means disease of the path between the cortex and the nucleus, or disease of the cortex itself. Hence the latter is sometimes called cerebral facial palsy, but this is not an accurate term, because this form may result from disease in the crus, or in the upper part of the pons. A more accurate name, therefore, and one by no means inconvenient, is *supra nuclear* palsy. On the other hand, the general palsy of the face may be either *nuclear* or *infra-nuclear*. It is often called "peripheral," but here again we have a word which it is wise to avoid. If it does not mislead, it is liable to distort a student's first conceptions. Such distortion involves a waste of mental effort, since it has to be rectified. The "periphery" properly means the distal extremity; "peripheral palsy" should be due to disease of nerve-endings—such paralysis as many poisons produce, not rarely, in other parts. But the word is not used in that sense in reference to the face. It is used to mean disease of the fibres anywhere in the course of the facial nerve, and even palsy that is produced by a lesion of the nucleus within the pons; that is of the central origin of the nerve, is sometimes spoken of as "peripheral."

This point,—the use of the word "peripheral" in connection with palsy of the face, may excuse a digression for a moment, to note an illustration of that oscillation of opinion that always attends advancing knowledge. The truth of yesterday may be untrue to-day and true again to-morrow. The grounds of an induction may become insecure when we know the facts more thoroughly, and yet, again, new facts may re-establish that which seemed exploded. When many of those who now teach, first studied medicine, they were taught that the nerve endings are the seat of the lesion in the commonest forms of facial palsy—that which follows exposure of the side of the head to cold. The cold, acting on the surface, was assumed to act on the nerve endings in the muscles of the face and thus to cause the paralysis. The same explanation was given of other palsies following an exposure to cold, such as the local atrophic palsy of children, "infantile paralysis." This pathology was purely hypothetical; no evidence of it had been ascertained, and there was no demonstrable analogy that could be produced to support it. Indeed, it now seems to us strange that the opinion should have been accepted so generally, since a little consideration will show how great a difficulty there must always have been in adjusting it to the facts. Why should cold act upon the extremities of one nerve, upon all of them up to the middle line, and no further, not a single muscle or fibre on the other side? Why should the effect of cold acting on the surface be absolutely limited to the endings of one nerve? This difficulty does not seem ever to have been recognised. When increasing precision of observation, and a wider comparison of facts, suggested irresistibly that such a paralysis must be due to a process acting on the fibres where all are near together, or on the contiguous structures from which they come, and when it was discovered that the latter was the true pathology of infantile palsy, the idea that disease of nerve endings was a cause of paralysis disappeared. All such palsies were ascribed to disease of the centre or nerve trunk. Again, new facts came, and have compelled a partial reversion to the old opinion, that a primary affection of the terminal parts of nerves is a frequent cause of palsy. But it is chiefly in the limbs that we find evidence of this. We have, indeed, returned to this opinion in some cases of paralysis of the face, but we have not returned to it as an explanation of the palsy that is limited to one side of the face, and for which it was once considered the adequate explanation. It is very unlikely, however the pathological pendulum may swing under the impulses of new discoveries, that we shall ever find reason to think that such palsy as you see before you in this child, is ever the result of disease of the nerve endings. We shall presently note the significance of this.

Let us return to the two main forms and their distinctions—the infra-nuclear palsy which is often complete, and the supra-nuclear palsy that is never complete. You perhaps know the hypothesis by which the difference is explained; it is indeed somewhat more than a mere hypothesis. In proportion as muscles act together on two sides, the muscle on each side is represented in both hemispheres of the brain; that is, either hemisphere can act on the muscles of both sides. Hence, in disease of one hemisphere of the cortical centre, or the path from that centre, the muscles escape in proportion to their bilateral association. When I say they escape, I ought to say they either escape, or they suffer only at first. They may be affected for a few days, but the initial weakness soon passes away, at least in great measure. Apparently the structural arrangements by which the hemisphere acts on those muscles on its own side are only slightly in habitual use, but their functional capacity quickly increases. This point, however, is one that we must consider another day; it would lead us too far from our present subject. It is the *complete* palsy—complete in range—of which an instance is before us, with which we are now concerned.

We can carry the diagnosis beyond that of the exclusion of supra-nuclear disease. Paralysis of all parts of the face, if it exists alone and is on one side only, means disease of the nerve after it has left the cranial cavity. Theoretically it does not; the whole nerve *might* be affected alone within the skull. But, as a matter of experience, it is not. The statement I have made is true in fact. Further, if there is no indication of a cause acting on the nerve after it has emerged from the canal in which it passes through the bone, such a paralysis means disease of the nerve within this canal.

This conclusion is, I need not say, of the utmost practical importance. In a large number of cases of facial palsy —in the majority indeed—it takes us at once to the seat of the disease. But to be able to use such a diagnostic rule you must know its reasons. Without a knowledge of the reasons few such rules are useful, because you cannot feel so sure of their validity as you must feel if you are to employ them with the confidence that is essential. You cannot feel sure that there are no exceptions, unless you realise why there cannot be exceptions. The facts on which this rule is based are these:—First, there are two portions of the nerve which cannot suffer alone on one side only: these parts are the beginning of the fibres and their terminations—the cells from which they spring, and the structure in which they end within the muscle. These parts are so situated that all of those of one nerve cannot be affected on one side and alone. The nerve cells occupy a considerable area of the pons, and those for the muscles of the eyelids are some distance from those for the lips. The cells are adjacent to other important structures. Hence, the whole of the nucleus cannot be damaged by one lesion, so as to affect all parts of the face without other structures suffering, and suffering in such a manner as to give rise to obtrusive symptoms. Moreover, this is true not only of the nucleus, but also of the fibres proceeding from it during their passage through the pons. All the fibres of the nerve are never damaged within the pons without adjacent structures being also damaged so as to cause conspicuous symptoms. Thus the isolation of the paralysis of the face, the affection of the whole of one nerve and of that only, excludes an organic lesion within the pons.

The nerve-cells may suffer, and may have their function abolished apart from what is commonly termed an organic lesion. They may cease to act in consequence of a degenerative process, or in consequence of the influence of a toxic agent. This is true, also, as I have already intimated to you, of the other extremity of nerve fibres—their termination within the muscles. These are now known to be occasionally the seat of degenerative processes, and they are also known to be the parts influenced by certain poisons. But such agents do not act only on the nerve endings of one side. The reason for this is important, and, as I just now said, must be specially noted. It is a general law of extreme importance. The same structures on the two sides possess the same pathological susceptibilities. They suffer together in degenerative processes, and they suffer together under toxic influences, whether these be what we commonly term poisons or whether they are only seen in what we call diseases. Thus both sides of the face may be paralysed throughout as part of diphtheritic paralysis, and as part of multiple neuritis, due to some toxic agency,—but one side of the face never suffers alone. This is the point to which I adverted in speaking of the theory that cold acts on the extremity of the fibres. Our present knowledge shows that it could not affect the terminations of all the nerves on one side and none of those on the other. Such a cause must, if it acts on the extremities of the nerves, cause bilateral paralysis. As a fact, it does so. Remember, then, in all parts of the nervous system, and in relation to all cases of the character I have mentioned—those due to degenerative processes and to toxic influences—the effect is bilateral; and a complete palsy of any part on one side practically excludes such general causes.

The law which underlies these facts, therefore, may be expressed thus: A palsy which is directly due to a *general* cause is bilateral; conversely, a bilateral palsy indicates a general constitutional influence. A unilateral palsy is due to a local cause; it is not the direct result of a general process. I use the word "direct" result in order to be strictly accurate. The meaning of the restriction is this. The direct effects of general causes are due to their action on the nerve elements themselves; but general states sometimes cause local effects by acting on the vascular system, or they may predispose all parts to suffer from an influence which acts only on one. Moreover, a unilateral palsy means a process which begins outside the nerve elements and affects them secondarily. That is a most important general law, applicable to all parts of the nervous system—most important to remember in all circumstances and in all localities.

Next, the limitation of the palsy excludes disease of the nerve within the skull. After it has left the pons, a process outside the nerve fibres cannot reach a considerable degree and affect the facial nerve alone. The auditory nerve is contiguous to it from the surface of the pons to the bottom of the internal auditory meatus, and a process external to the facial nerve, sufficiently severe to paralyse it completely and in all its parts, must affect the auditory nerve as well; often it affects other nerves that are in the neighborhood. On the other hand, a cause acting on the facial nerve after it has passed through the skull will be conspicuous; it is always an injury or a considerable local inflammation, and, in either case, is obvious. Hence, therefore, you can now understand the foundation of the important diagnostic rule which I laid down before: that a complete unilateral palsy of the face, without other symptoms, must mean disease of the nerve as it passes through the bone.

The diagnostic problem is thus narrowed in regard to the seat of the lesion, and this facilitates the process of diagnosis even more than may at first sight appear. We cannot, indeed, carry further the process of exact localisation, except that, if the short length of nerve which intervenes between the origin from it of the vidian above and the chorda tympani below, is affected, taste is lost on the side of the tongue in front. But this is of little real value, because a morbid process may begin at one spot and may spread through a considerable extent of the nerve. The state of the palate gives us no localising information. Paralysis of the palate is never produced by disease of the facial nerve. The belief that it is so is one of those curious pathological myths which have arisen from the misinterpretation of what may be termed normal abnormalities, if you will forgive the expression—deviations from perfect symmetry, which have no significance for us except in so far as they may lead us into error.

But the special aid in diagnosis which we derive from this restriction of a morbid process to the part of the nerve which lies in the narrow winding canal in which it passes through the temporal bone, is due to the aid it gives in the next step in diagnosis—the step which is of chief practical importance, the *nature* of the lesion of the nerve. In this part of its course the nerve suffers from only three morbid processes, and of these three one can be excluded without difficulty. The processes are:—(1) primary inflammation of the sheath and interstitial tissue, which is the cause by which the most common form is produced, that which is due to cold; (2) the spread to the nerve of inflammation due to ear-disease; (3) its compression or destruction in consequence of a growth in the bone. It is the last of these which may safely be put on one side unless there are other obtrusive indications of a growth, because it is a very rare cause, and not to be thought of unless such indications exist.

The diagnosis between the two forms of inflammation, that which is communicated from the ear, and that which is primary (the so-called "rheumatic" form), is seldom difficult. They differ in the period of life at which they chiefly occur; although each may be met with at any age, ear-disease is a rare cause except in childhood, while primary neuritis is seldom met with till childhood is over. Moreover, the mischief in the ear which spreads to the nerve is generally considerable in degree and in duration. In most cases there is actual disease of the bone, and it is the progress of the caries that brings the associated inflammation into such proximity to the nerve that this becomes affected. In such disease there is almost invariably perforation of the tympanum and a constant discharge from the ear. If there is no history of discharge, disease of the ear is not likely to be the cause. But this rule, true of facial paralysis in general, is not invariably true of it. Exceptions are met with, and this case is an illustration of the fact. We often hear vaguely of exceptions which "prove" or "test" the rule. The rules to which this saying is applicable are in most cases general rules; they are laws that are true of the majority of instances which come under them; if these are separated and scrutinised, it will be found that the exceptions occur under special conditions, and that these special conditions have been ignored in formulating the general statement. We have to be chiefly guided by the majority of cases, but we should recognise the existence of exceptions, and know when they may occur, that we may search for their indications if we have any reason to suspect an exception to our rule. If we attempt always to give weight to them, such weight as we give to the majority, we shall be in constant uncertainty. Indeed, it is not unlikely that we may come to the conclusion, as a distinguished scientific man remarked to me of the impression left on him by one of the most famous teachers of his early days, that "no sane man could make a diagnosis."

The exception to the rule that obtrusive signs of caries long precede the facial paralysis which results from otitis, depends upon the anatomical conditions of the ear. This has not indeed been actually proved, but we know that exceptional conditions often exist; they explain that which would be otherwise inexplicable, and which nothing else explains. In some cases, fortunately not common, the facial nerve is separated from the tympanic cavity by a layer of bone so thin that inflammation can readily pass from the cavity to the nerve sheath. In such, bone disease is not necessary for the extension of inflammation to the nerve.

You can now understand why I asked so carefully about the earache. It is not enough to ask if there has been discharge. Discharge generally means disease of bone: disease of bone is the common cause of secondary facial paralysis, but the nerve is sometimes affected by extension when there is no bone disease. This case seems to be of that character, and hence, I am anxious to impress its facts upon you. It is an illustration of one of the two chief simple sources of error in diagnosis, not seeing the common, not knowing the rare. I say simple causes as opposed to the more complex sources of error in reasoning.

This is an instance of the rare; it is an example of the occurrence of facial neuritis by extension from the middle ear, without bone disease. It is exceptional, because it must depend on exceptional conditions. A "passage" of inflammation implies a way for it to pass out. Normally, there is no way free enough for the passage of simple inflammation. But simple inflammation may spread from the lining membrane of the tympanum to the nerve, when the layer of bone which separates the tympanic cavity and the nerve is thin. Indeed not only may it be thin, it may be even actually deficient. Inflammation not only may, but, if considerable, must then spread to the nerve. Vessels also pass through the bone, and by these an intense inflammation may pass; but it is doubtful whether a simple catarrhal inflammation, such as this child seems to have had, would do so, if the conditions were normal.

One other point, and that of great importance, the case also illustrates. The earache of children is almost always due to inflammation of the middle ear. Most of you know its character from the recollections of childhood, for few children escape some attack, and the peculiarity of the pain impresses itself on the memory. It is one of the maladies that may be called "domestic diseases"; few mothers dream of sending for medical assistance for what they are pleased to call "simple earache." But between the earache which lasts a few hours and then passes off, and the earache which is the prelude to a suppurative inflamation, there is every gradation. There seems to be no difference in the character of the pain.

It is important to remember this. It is one of the facts that should be part of the education of the mother—one of the many facts that might, with great advantage, replace the rubbish that has drifted into the maternal mind: by no means harmless in the minds of the many persons who "get on much better without a doctor." I lately saw a child with infantile palsy of one leg. This leg had suddenly become powerless. The mother—with no excuse of poverty or station—consulted her female friends, and, resting content with their assurance that it was "only the teething," and that "the leg would soon get all right again," allowed five whole months of utter immobility to pass before she thought it necessary to ask her doctor to look at the child's leg. And many cases of suppuration in the ear and of disease of the bone are due to neglect of the warning of simple earache.

In the case of children who are liable to earache, great care should be taken to guard against exposure of this part of the head to cold, and especially to cold east winds; and still greater care should be taken to get rid of an attack as soon as possible. How seldom are the lessons learned that are, nevertheless, familiar in proverbial form, such as that embodied in the adage about the "stitch in time." We could wish the statement were literally true, and that the absence of early treatment, which so often would cut short a grave disease, could be made up for by the amount represented by the multiplier of the proverb.

If you take all the nerves of the body, and consider the frequency in which they are so diseased as to cause symptoms, I think the fifth, the sciatic, the ocular, and the facial nerve would be their order. The reason for the frequency of facial paralysis is not yet entirely understood. We can understand, however, why it is so obtrusive when it does occur. It is manifested with peculiar readiness, because outward swelling is prevented by the rigid walls of the canal. Hence the inflammatory effusion compresses the nerve fibres, and at once interrupts the conduction of the motor impulses, quickly causing inability to move the facial muscles. But this, although it explains the fact that even a slight inflammation causes considerable palsy, leaves its actual frequency still mysterious. Paralysis of this nerve from cold is certainly more common than we should anticipate, considering that other nerves are not less exposed. Perhaps the cause is to be found in some conditions of the circulation within the canal, in consequence of which a congestion that would otherwise be transient and harmless, leads to undue stasis, and becomes an actual inflammation with all its grave results.

The more remote causes of the primary neuritis are obscure. It is occasionally associated with the diathesis that causes fibrous and muscular rheumatism, and which is probably not far removed from that which causes or results from gout. I have twice known a patient to have facial neuritis at one time, and at another an analogous rheumatic inflammation of all the nerves at the back of the orbit. Remember, too, that what we call fibrous rheumatism is probably also not very different from an inflammation. Certain it is that this still mysterious muscular rheumatism may become inflammation. Many cases of sciatic neuritis, certain and severe, arise by the traceable extension, along the fasciæ to the sciatic notch, of a primary lumbago.

We will not attempt to ascertain, in this child, the electrical irritability of the muscles. We know what condition we should find. No complete paralysis of a motor nerve continues for a month without the nerve fibres degenerating below the lesion, and there is always loss of all irritability of the nerve trunk, loss of the faradic irritability of the muscles, and increased irritability of the muscular fibres to voltasim. We should learn nothing by the examination, and you have frequent opportunities for observing the facts. Without doing good to ourselves, we may do harm to the patient. The paralysed muscles will need electrical stimulation, and in the case of children it is necessary to be extremely careful in the application of electricity. If you so use it as to cause pain the child will be frightened, and will dread each therapeutical application. Once thoroughly frightened, a child seldom loses the dread, and no child can endure a frequent distressing emotion without harm. If care is taken, all that is needed, or almost all, can be achieved without the production of any of this injurious alarm. But to secure this result the first application must be so feeble that no new sensation is produced. Indeed, it is well, the first time, not to allow the current to be strong enough to be felt. Then, if the current is gradually increased in strength in successive applications, after a few days a strength may be used that will make the muscles contract visibly without eliciting a tear. It is surprising how strong a current children will bear if this plan is adopted. For the same reason you should never use the faradic "*current*" in the case of young children. Move the hammer with your fingers instead of permitting it to oscillate automatically. The less frequent momentary currents will cause a muscle to contract with merely a peculiar shock-like sensation, devoid of pain. If you use the repeatedly recurring shocks, produced by the automatic interruption, a painful stimulation of the sensory fibres is produced, even with a lower current than will cause the muscles to contract. The difference, in the case of children, is of great importance, for the reason I have mentioned. For the same reason, also, if you find it impossible to cause visible contraction of the muscles without producing distress, be content with a little weaker current. I cannot understand the prevalence of the notion that in any therapeutic procedure, with drugs, electricity, or any other agency, no good is done unless a certain physiological effect is produced. Surely the production of a manifest effect is only a question of degree. If a certain strength of current cause a visible contraction of a muscle, it does so by making a large number of the fibres contract with sufficient energy for the effect to be seen. Are we to assume that a current just below this strength causes no contraction because we cannot see the whole muscle shorten? There must be contraction of the fibres before we reach that degree which causes visible movement, and it is surely impossible that this stimulation can be without the influence on the nutrition of the fibres, and on the maintenance of their irritability, which the stronger current has. On the contrary, the consideration of the fact suggests distinctly that such a weaker current only needs to be continued for a longer time to do all that the stronger current can do. Remember, all that electricity can do in such cases is that which I have mentioned. We have no reason to believe that it has any influence on the nerve fibres; it does seem to keep the muscles in a better condition for the nerve impulses to act upon them when conduction is restored, and to enable them to respond better to the motor influences when these can again reach them. Hence, in all cases in which the paralysis is complete, or considerable, for more than two or three weeks, it is desirable to employ electricity.

Moreover, we do not here need an electrical examination to assist in forming a prognosis. This is, it is true, one of the most important services electricity renders in such cases. It is surprising to those who cannot read its language, that of two cases of facial palsy alike complete and of the same duration, say two weeks, an electrical examination should enable a positive statement to be made that one case will be well in a fortnight, and that another will endure for months and will never completely pass away. And yet it is so. But we have other aids in prognosis. We can often draw conclusions of great value from the simple consideration of the conditions under which a malady developed, and, when a process in one place is due to extension from the same process elsewhere, the character of the primary symptoms is sometimes of significance. Here we have the fact that the paralysis resulted from an inflammation which was transient. The pain in the ear lasted only a few hours. Doubtless the process may have lasted longer, since when effusion and swelling occur, the pain of inflammation often lessens, but the duration of pain which is due to inflammation in the ear is of some value as a guide to the duration, and, therefore, to the intensity, of the process. Thus judged, the process here was brief, and, being brief, was also slight; for duration and degree are proportioned in acute inflammation. From this we may conclude that the inflammation of the nerve was probably only such a brief moderate process as would be limited to the sheath to which it first extended. The effects have already lasted a month, but before long some signs of improvement may be expected to occur. Over and above this, we have also the general law that cases of facial paralysis from ear disease, on the whole, run a more favorable course than cases of facial paralysis from the primary neuritis. The reason for the average slighter severity of the affection is, perhaps, to be found in a more limited extent of the inflammation. It is reasonable to assume that an inflammation which is communicated will not affect so considerable an extent of the nerve as one which is due to a primary process.

In all such cases of facial paralysis the treatment must be twofold. Besides that suited to any recognisable constitutional state (which I assume you do not need to be told) we have to treat the primary disease of the ear, and we have to treat its consequence. In the case of ear disease, this must first engage our attention, but the measures needful do not come within my own province to describe in detail. The most important, however, is unquestionably to afford a free exit to any pus which may have accumulated in the cavity of the tympanum, and to guard against any obstruction of such exit, including that which is produced by non-absorbent cotton wool. Absorbent cotton wool changed two, three, or six times a day, according to the amount of discharge, is safe. The importance of cleansing by antiseptic washes you will have already learned. For the secondary inflammation of the nerve sheath we can do nothing directly, except by the application of counter-irritation. This is, indeed, the chief local measure in all cases, whether produced by cold or produced by extension. A blister should be applied over the mastoid process, and should be repeated as soon as the skin will permit. Never put a blister in front of the ear, over the place of exit of the nerve. You cannot blister without producing a little subcutaneous cellulitis, and I have known a trifling cellulitis in this situation to be a cause of facial paralysis—the inflammation reaching the nerve sheath, and spreading along it into the Fallopian canal. In all recent cases, moreover, hot fomentations applied over the ear and its neighborhood, constitute a very useful measure. They should be used for one quarter of every hour during the first day. We do not precisely know how hot fomentations act, but it is certain that no measure has so potent an influence on inflammation in its earliest stage. Probably the stimulation of the sensory nerves and the heat combined cause alternate contraction and dilatation of the vessels, which lessens the blood stasis. Perhaps, moreover, the nerve stimulation has some even more direct influence on the process of inflammation. But whatever the explanation, the fact is certain.

The last element in the treatment is the application of electricity. This is only needed when the nerve fibres undergo degeneration. As long as the nerve retains any degree of excitability, and as long as the muscles contract when the induced or faradic current is applied to them, electricity is not needed. Its use is to keep up the nutrition of the muscles and to keep up their excitability. We have conclusive proof that it does the latter; we have no proof that it does the former, because we do not find that the muscles to which it is applied waste less or waste more slowly than those to which it is not applied. But it is certain that functional excitability cannot be maintained without nutrition being also influenced. You know the kind of electricity that should be applied. This is one of the elementary facts which every student learns or should learn as his first acquisition in the therapeutics of the nervous system. You must apply the kind of electricity to which the muscles respond, and this, as we have seen, is the voltaic current. You apply it in such a manner that it is interrupted, not frequently as the induced current is, but slowly, as by *stroking* down the muscle with one electrode, the positive or the negative, to whichever the muscle is most sensitive. It is not always the same, even when the nerves are degenerated. We interrupt slowly because, as far as we can judge, the element in the form of electricity which causes the difference in the reaction is one of time. Muscular tissue is much lower in the scale of excitable tissues than is nerve tissue, and seems to be unable to respond to an electric current unless this has a duration considerably exceeding the very small fraction of a second which elapses before the automatic interruption stops the current of the induction coil.

LECTURE VII.

FACIAL CONTRACTION AFTER PALSY.

Gentlemen:—When I asked your attention to the chief symptoms of facial paralysis, I had to leave for subsequent consideration the troublesome condition which accompanies return of power—the secondary state of contraction. It is not a trifling matter in many cases. It may be almost as great an evil as the original paralysis. It is, moreover, a subject of considerable interest, and its study throws light on other conditions.

Observe carefully the face of the patient before you. The left side presents a slightly deeper naso-labial furrow than there is upon the right side. It would be natural for an observer to assume that the side on which the furrow is least marked is the weaker side, is that which was paralysed. As a matter of fact, such an assumption is commonly made by students in such cases. The furrows of the face are determined by muscular action, and where furrows are less there is less muscular action, and therefore muscular weakness seems to exist. The impression is strengthened by the aspect of a gentle smile; it is not easy to obtain it—the attempt often results in no movement, or in one that is much in excess of "gentle." But the request for one, cautiously put, often obtains what is desired. You see it does so here. You see that the gentle movement is distinctly greater on the side of the deeper furrow, and the smile thus seems to confirm the opinion that there is more power of movement on this side.

But our ability to excite emotion may be sufficient to make a patient laugh; or we may obtain an energetic movement in elevation of the upper lip. In this you can see how much slighter the strong movement is on the side on which the slight movement was greatest. On the side of the deep furrow a slight movement is greater, but a strong movement is much less. Note, also, that when the eyes are tightly closed there is the same difference. An excessive degree of slight movement is associated with a diminished degree of strong movement.

But you may have already observed for yourselves that there is more than this difference in degree. Even in the gentle smile you could not fail to see that the eyelids were more approximated on the side on which the slight movement was most considerable, and that this was conspicuous also with the stronger movement. It is very marked in the laugh. You know that emotional movement normally involves all parts of the face. The ocular fissure is narrowed in a laugh, the eyes may even be almost closed, while the angles of the mouth are drawn outward and upward by the zygomatic muscles in the ungraceful distortion of the face which custom makes us appreciate so keenly. This associated action is increased in the condition we are considering. With the increased readiness of action there is an increase in the associated action of the different parts. Moreover, with these there is associated the feature to which I first directed your attention—the muscular contraction which deepens the furrows. This muscular contraction is a persistent state; it is accompanied by, and, indeed, I might say, illustrated and emphasised by occasional slight clonic contraction in the muscles, which you will have no difficulty in perceiving in this patient, if you presently carefully observe her.

The furrows which come with years, sooner or later, are the result of the shortening of the muscles which have been most frequently employed in their varied uses, and especially in the expression of emotion, a shortening increasing with years and being increased in effect by the loss of elasticity of the skin. The emotions that are chiefly dominant impress their transient influence on the muscles by long repetition as a permanent effect. Everyone knows the varying expression which the furrows of age produce in the aspect of the face, and those who know the story of the life can often read it in the lines. Yet most persons remember instances which cannot altogether be thus explained. We sometimes meet with a face which bears the aspect of sorrow as the sequel to a tranquil life, or see the furrows of a constant frown in one who is seldom angry, or the transverse frontal lines when surprise or concern has seldom been dominant. Of course, many of the lines which Time's fingers trace are due to that which is common to all persons, habitual movements which have only a little to do with emotion. The zygomatic muscles, for instance, are constantly employed otherwise than in the smile or in expressing pain. Their contraction is the result of their habitual action, defectively opposed. But when all allowance is made for such influences, there remains much contraction which we cannot explain except as the result of hereditary influence—the effect of emotion in past generations, persisting to the present, and telling a story which is untrue of the present, but, like so much else that is mysterious, is true of a perhaps distant past. It seems to be an indication of the continuity which makes our race extend unbroken, in all its variations, through time which we cannot estimate.

The fact that the contracture which follows facial paralysis resembles in its effects the condition which age induces, renders its effect most distressing to the young. The contracture follows inevitably in every case of facial paralysis, with the exception of those which last but a few days. If the paralysis is moderate, the contraction is moderate; if severe, so is this sequel. If the time of life has come when lines are proper, then the contraction reproduces the due symmetry of the two sides of the face; and if it does a little more, the excess is almost imperceptible. Although the associated overaction may cause a difference in the expression of emotion, unless there is energetic movement, the difference is slight. But that which is natural in age is unnatural in youth. The contraction produces its furrows in all periods of life, and the result in youth is that a face may one side correspond to the proper aspect of "sweet seventeen," and on the other to forty-five. The contrast is an evil as great as is the facial paralysis itself, and even greater. In the young, during the stage of palsy when the features are at rest, no difference is perceptible between the two sides, and it is only on movement that distortion occurs. But this late contraction causes a conspicuous contrast between the two sides, which is constant at rest. It is the more distressing because no hope can be held out that it will pass away. So keenly is it felt, that one patient of mine, a girl of eighteen, formerly epileptic and not of very stable mind, was so distressed at the daily reflection in the looking-glass, which she could not avoid, that she one day wrote a letter to me. It was to thank me for my efforts to do her good, but to say that she could bear it no longer, and that when the letter reached me she would be dead. The letter was to reach me on my visit to the hospital, and when it reached me it was true.

In the old, however, the contraction is cosmetic in its influence. The early palsy obliterates the bilateral symmetry, because it substitutes the smoothness of youth for the wrinkles of age, but the symmetry is restored by the late contraction. It is curious, by-the-by, to note how difficult it is for many persons who have long ceased to be young, to believe that the normal side of the face is not that which is affected. In the early stage of the paralysis the smoothness of the affected side causes the furrows on the other, familiar as they are, to seem by contrast entirely strange. That which is so unseemly cannot be natural. They assume, therefore, that the side toward which the mouth is "drawn," is that which is affected. The conviction is not, as you might think, confined to one sex; I will not say that it is not a little more common in one, but it is frequent enough in the other. A mirror has an interest for men as well as women.

This late contraction, as I have told you, always follows facial paralysis except when very slight. It accompanies return of power, but only return of power that is incomplete. This statement may surprise you. It implies that recovery of power is incomplete except in the slightest cases. You may have seen many cases of paralysis of the face which seem to have perfectly recovered. But you will find, on careful examination of the movements, that recovery is hardly ever perfect. You will find, also, that when you find any indication of contraction and associated overaction, you can always detect imperfect power, however long it may be since the acute affection. Indeed, the overaction may be more conspicuous than the persistent defect; but the defect is always there, although it may be inconspicuous. In every organ and structure of the body the condition after inflammation is never as it was before. Could we perceive the minuter structure, we should be astonished at the degree of persistent change. In most parts the effect on function is not perceptible, because the function of the organ is manifested by the combined action of all its parts. Where each part has a separate function, the effect of persistent change is evident. This is so with the face. The evidence of imperfection is the greater the more pronounced the function, and the more delicate its adjustment. The face is the great vehicle of emotional

expression. No pain or sense of pleasure comes and goes
without a change in the state of contraction of the facial
muscles, too slight perhaps to be analysed, but too definite
to be unperceived. It is not surprising that an organ of
expression so delicate should manifest the slightest interference with the process of conduction of the impulses
from the brain.

This peculiar condition does not cause suffering. It
may be attended with a sense of tightness in the face at the
onset, but afterwards the patient is scarcely conscious of
it, except when movement brings its presence to the mind.
This, however, is not more than the patient quickly gets
used to. The muscles in the lower part of the face on one
side are, in general, adequately opposed by the action of
those on the other side. If the second side should subsequently become paralysed, the contracture instantly
becomes far more obtrusive both to the patient and to
others. It is indeed very rare for the second side to suffer
at a subsequent period, and very few instances are on
record. A case lately came under my notice : A lady had
been under my care five years ago for severe paralysis of
the right side of the face, which had left considerable contracture. One day she was startled by finding a great
increase in the contraction, and that there was deviation of
the mouth considerable even on rest, greatly increased by
movement. On looking in the glass, her previous experience made it at once obvious to her that the left side had
become paralysed. She came to London in a day or two,
and the condition was very striking; the vast increase in
the effect of the old contraction, consequent on the loss of
power on the other side, caused extreme distortion. Fortunately the attack proved brief, and in a fortnight the fresh
palsy had passed away, and her state was as before. How
far the quick recovery was due to the treatment adopted
(which I have described to you before) or to the slightness

of the inflammation, I cannot say. Remember that this is one of the maladies in which nine-tenths of that which can be done by treatment, can be done only during the first few days. It is one of the many diseases in which therapeutic knowledge on the part of the general practitioner, perfectly ready for use, is of the utmost importance. Physicians are impressed perhaps unduly with this fact, because most of the patients who have reason to seek their subsequent aid, are cases in which the needed treatment has not been forthcoming. It is impossible to overlook the room there is for more knowledge of the proper treatment of many diseases, knowledge which is such as to enable the necessary measures to be at once and confidently adopted. Practitioners, perhaps, do not adequately realize how large is the proportion of all possible good which they alone can do.

I have dwelt at length on the features of this late contraction because, trifling as it may seem to be, its practical importance is very great. As I said, although the annoyance occasioned by the paralysis may be greater in degree, it is almost always limited in its duration. That which is caused by the contracture is generally considerable in the first half of life, and it is not limited in its duration. It only ceases to annoy when the effect of years neutralizes, by balancing, its effect. Why it should be thus persistent will be seen if we endeavour to perceive something of its cause.

Its cause is certainly the functional state of the nucleus of the nerve produced during the period of palsy. During the state of immobility a condition of the motor centre of the nerve has been induced by which a slight degree of excitation causes undue action, although the hindrance to conduction along the nerve is such as to lessen the total amount of energy which can be transmitted. It is not difficult to conceive that the narrowed fibres should have their capacity for conduction restored so that they conduct perfectly a slight amount of nerve energy, and yet the amount of force they can conduct is permanently reduced. There is an explanation of this in the mode of conduction; but I cannot now make it clear, because I should have to begin far, far back, in the elements of molecular physics.

This state which follows paralysis is evidently due to a state of the motor elements of the nucleus, that is, the structures from which nerve energy is evolved. These act too readily, they act spontaneously so as to cause the twitching, and they act on each other too readily, so as to produce the associated overaction. The structures from which proceed the fibres for the different parts of the face, are connected in the nucleus, and these connections which subserve the normal associated action present in health a certain resistance, which limits this action. It is on this resistance, developed by normal function, that the due relation of the associated action depends. That resistance becomes lessened during the palsy. It is lessened in consequence of overaction of the centre. Wherever we have overaction of nerve structures extending from one to another, the resistance between them is lessened, and the activity tends to increase the condition. It is as if the grooves in which wheels run are deepened into ruts, and the continued motion of the wheels precludes any diminution in the depth of the ruts. It is as if a channel for water had become widened by the flow, and yet is inaccessible to any direct process of repair; the constant flow of water through it precludes any natural reproduction of the natural resistance. So this overaction of the nerve centre, once set up, goes on.

It is not difficult to perceive how it is probably set up. At least it is easy to see a mechanism that must inevitably have the result. Normally, the action of the central structures of a motor nerve is regulated and restrained by impulses which reach the centre from the contracting fibres of the muscle by the afferent nerves, stimulated by the contraction of these muscular fibres. The voluntary and other impulses continually act on and excite the nucleus when the nerve is interrupted. In the absence of the control which should be exerted on them by the afferent nerves from the muscles, this must lead to overaction of the nerve centre. Moreover, the obstruction in the nerve fibres cannot be without its influence in disordering the centre. Nerve fibres conduct by a process similar to, but less in degree than, that by which the nerve force is generated, and although we do not understand the process, we must conceive that the arrest of the transmission onward of the energy has an effect on the functional state of the centre. If the molecular changes of conduction cannot pass down the nerve, this passage into it must, after a time, be hindered, although the production force is not lessened. Hence also the abnormal action within the centre is increased.

The occurrence of spontaneous twitching in the face is explained by the tendency to overaction on the part of the motor structures in the nucleus, and the fact that slight impulses of nerve force can pass readily through the nerve.

I should like you to note the difference which there is between this form of contraction after paralysis and that which you meet with in the limbs. In these, the loss of power with changed reaction to electricity is often followed by muscular contraction, but the contraction is in the opponents of the muscles that are paralysed, and not in these muscles themselves. It is due to the fact that the diminished extension permits shortening. But the peculiarity of the contraction after facial paralysis is that it occurs in the muscles that are paralysed. The difference must be referred to the peculiar conditions which cause the loss of power. Nowhere else do we meet with so absolute an arrest of conduction in consequence of a lesion of the nerve which afterward lessens or almost recovers, as we do in the case of the facial nerve. It is, of course, due to the course of the nerve within the bony canal. Moreover, nowhere else have we the constant energetic influences from the brain exciting the motor centre, as in the case of the centre for the face. I have already dwelt on the peculiar relations of the face to emotion, which involve the constant excitation of the centre. The perfect bilateral association of the action of the two sides is another condition which precludes any cessation of the action of the brain upon it. It is very different with the spinal centres for the muscles which may be paralysed by disease of the nerves. Thus we can understand that the unique conditions of function and disease, in the case of facial paralysis, give rise to a condition here which we do not meet with elsewhere.

You will perceive from what I have said that the condition is beyond the influence of treatment. I have never been able to form an opinion on the question whether it lessens with time or not. If it does, the diminution is very slight. But there are three practical points connected with it. The first is that its advent is certain in all cases in which faradic irritability is lost or considerably reduced. Except in the very rare cases of absolute lasting atonic palsy, some power is always regained, but with it comes this sequel. Since it is annoying, it is always well to warn your patient that when the face recovers power, it will tend to overact, and that this is inevitable. Otherwise the patient is apt to think a fresh morbid state is coming on. When it begins, I think it well to stop electrical treatment. Its commencement coincides with, or soon follows, the return of voluntary power. There is then enough power to influence the nutrition of the muscular fibres, and to render electrical excitation unnecessary. On the other hand, the stimulation of the sensory nerves by electricity tends, I think, by reflex action, to increase the contraction. The third point is the only measure which has, or can have, any influence. A trifling effect on it is probably exerted by what may be called "downward massage" of the face. The fingers should be drawn from the zygoma to the angle of the mouth repeatedly, with gentle pressure, for a minute or two, several times daily.

It is not pleasant to be able to advise so little. But, Gentlemen, remember this—it is true of ourselves and it is true of our patients—next to knowing what can be done and how to do it, the most important thing is to know what cannot be done. Sad, indeed, is the waste of time and money caused by efforts to get that which cannot be; sadder still is the waste of hope—hope created by baseless expectation—and the destruction of it that we call disappointment, disappointment that would not be were it not for the anticipations that have no justification.

LECTURE VIII.

ACUTE ASCENDING MYELITIS.

Gentlemen:—Before we proceed to the special subject which we are to consider to-day, I propose to describe to you a case I saw last evening. There are many diseases which, because they are rare, or for other reasons, seldom come under the student's notice, and although description can never take the place of personal observation, it may be useful if the latter is impossible, and it may be the more useful if a case has been recently seen, and its facts are fresh and vivid in the narrator's mind.

The patient whom I saw (with Dr. Parnell and Dr. Grayling, of Forest Hill) was a lad of 19, paralysed almost completely from head to foot, pale, with the distressing struggle for breath that comes when breathing power is getting less and less. The story of the case is as follows:—

A month ago he had his first lapse from virtue, and the consequence was—not a rare one—an attack of gonorrhœa; it was treated, and he recovered in a fortnight. He had not previously been in good health, having been overworked at night. A little more than three weeks after the onset of the gonorrhœa—about a week after its termination and six days before I saw him—his legs, one day, became weak, he stumbled on the stairs and in a few hours the weakness became considerable. The next day his abdominal muscles were feeble, and the legs powerless, but with excessive knee-jerks (which disappeared two days later). On the following day his arms began to lose strength. This weakness increased and extended during the next three days, and he became febrile. When I saw him last night his state was this: He was sitting, half upright, in a chair, breathing, as I have said, with difficulty, by means of his diaphragm and lower intercostal muscles, at the rate of 56 respirations per minute, the deficiency in quantity of air inhaled having to be made up by increased frequency of respiration. His temperature was raised to 103°. His legs were absolutely powerless, without a trace of knee-jerk—flaccid palsy. From the right sole there was no reflex, but a touch on the left caused a considerable movement of the foot; there was no abdominal reflex, but a stroke on the skin left the bright red line, lasting a long time, which shows acute disturbance of vascular innervation. He could just flex his fingers, and that was all the movement left in his arms. The abdominal and trunk muscles were paralysed, except the lower third of the intercostals and the diaphragm. The upper part of his thorax was motionless, and the sterno-mastoids were also powerless. He could swallow, and he could just speak, although with difficulty. The left side of his face moved less than the right; the left masseter contracted a little less than the right; and movement of the eyes was accompanied with brief but wide nystagmus, greater on movement to the left side than to the right. Without any discoverable cause—without any hot application or the like—there had developed, within a few hours, a very large bulla over the inner side of the right ankle.

It was thus a case of acute ascending paralysis, using the words in a symptomatic sense, and it had naturally been regarded as a case of that mysterious disease to which the term is specially appropriated, and which is also called "Landry's paralysis." I think that this reasonable conclusion is very near the truth; but there was evidence of more than we have in that disease. Indeed, the case is particularly instructive from the point of diagnosis. "Acute ascending paralysis" depends apparently on a peculiar arrest of central, and perhaps peripheral function, by a peculiar blood state. In it there is no pyrexia; the loss of function presents a perfect bilateral symmetry, never transgressed by such a deviation as the preservation of the plantar reflex on one side when it is lost on the other. The course of the symptoms is uniform, the knee-jerk is lost from the first, and never excessive at the beginning, as in this case. Although it is said that some parts may be passed over in the upward march of the palsy, as, in this patient, were the lower intercostal muscles, yet I think it is doubtful whether this is really true of Landry's paralysis, and whether the assertion has not been due to cases—such as this—which may so readily be mistaken for that affection. The symptoms point to a derangement of the functions of the spinal cord, slightly but distinctly irregular, both in time and place. There is no doubt that Landry's paralysis, as I have just said, is the effect of a toxæmic state on the nerve functions; but all toxæmic conditions, in their isolated action, cause *symmetrical* arrest of function; and the difference we had here in different parts and at different times, showed that there was an irregularity in the cause of the symptoms,—something more irregular than the simple inhibition of a blood state. The early excess of the knee-jerk, together with the difference between the two plantar reflexes, and the nystagmus, pointed to a central affection; the first of these suggested a process in the spinal cord beginning, and at first most intense, above the lumbar enlargement; while the pyrexia, developing late in the course of the affection, was evidence of inflammation rather than of a primary blood state. Had the fever been due simply to the cause of the symptom, it would have occurred early. Its indication—that there was inflammation—agreed with the spinal symptoms. However set up, acute inflammation in the nerve centres always tends to more or less irregular and random influence. This is no doubt largely determined by the participation of the vessels in the process, and by other causes, such as a slightly greater intensity in one tract of grey matter in which it may spread, or the local influence of a small inflammatory extravasation. Hence inflammation, when set up, extends in a more or less random manner, and its symptoms are often irregular, at least in some degree.

The significance of these indications is that, in this patient, we had to deal with acute ascending myelitis. This is the condition between which and the special "acute ascending paralysis" there is generally most difficulty in diagnosis. Although the initial excess of the knee-jerk suggests that the inflammation began above the lumbar enlargement, and that the centres in the latter were merely irritated at that stage, yet the complete palsy of the legs shows, however, that even then the damage to the cord above the lumbar enlargement must have been considerable. As the inflammation spread, upwards and downwards, the damage to the centres in the grey matter of the lumbar enlargement caused absolute loss of the knee-jerk, but it would seem that, even to the last, the lowest part of the cord on the left side was not completely involved, so that a plantar reflex could still be obtained. Trophic disturbance, such as the large bulla on the side of the heel, does not occur in Landry's paralysis: it is an indication of an intensely irritative state of the nerves, transmitted to them from the spinal cord through the posterior roots. Although the posterior nerve roots conduct impressions upwards they transmit nutritional, influences downwards. This transmission is, indeed, probably by their own nutrition; the disorder in their molecules passes to the tissues in which they end. This bulla, in connection with the other symptom, showed that there was an intense irritative inflammation in the spinal cord. It was on the side on which the plantar reflex was abolished—that is, on the side on which the inflammation of the grey matter, in the lowest part of the cord, was most intense. The lad's condition did not allow of our ascertaining exactly the state of sensation.

But, in the lessened movement of the face, of the masseter, and of the eyeballs, trifling as all were, there was evidence that the inflammation was spreading upwards into the pons. It seemed to have passed the medulla, except that the heart-sounds were unnaturally short, with a peculiar character, seldom met with except in acute disturbance of innervation, as if the cardiac centre were becoming affected.

I have said that the opinion that he was suffering from Landry's paralysis was probably not far from the truth. There is no doubt that Landry's paralysis is due to an acute blood poison acting on the nerve centres, from below upward. All present knowledge leads us to ascribe an acute inflammation of similar course to a like cause. Both maladies occur under analogous conditions. It is probably only a question of a slight degree of difference in the nature of the blood poison, whether it simply arrests function and causes slight nutritional changes, or whether it sets up actual inflammation. Measles may be followed by typical Landry's paralysis, without evidence of inflammation; and measles may also be followed by an intense spreading myelitis. We are learning to see more and more distinctly the profound causal influence of toxic blood states in producing inflammation of the spinal cord as well as of the peripheral nerves, and it is most important to recognize the fact that allied blood states seem to cause one, or the other, or both. At present, multiple neuritis has the hold on professional thought which novelty always entails, and central diseases are sometimes overlooked. I have lately read some descriptions of acute fatal " multiple neuritis" (without autopsy), due to toxæmia, which I am sure were cases of ascending myelitis.

Among these blood states which act on the nervous system some of the most important are those connected with acute specific diseases—the diseases that are due to an organismal virus. The question was put to me in this case, Was the lad's malady connected with the gonorrhœa? I do not know that anything of the kind has been hitherto observed as a result of gonorrhœa, but a causal relation between the two is highly probable. We know how such states are produced by various acute specific diseases; and we know that the poison, the organized poison, of gonorrhœa may induce a subsequent toxæmic state which is manifested by the arthritis of "gonorrhœal rheumatism." Let us consider, for a moment, what this means.

The organisms of such diseases seem to cause these consequences, in most instances, by producing a chemical organic poison in the blood, and analogy makes it highly probable that such an agent is the cause of the arthritis. But we know that such products can act on the nervous system, peripheral and central, inducing grave disturbance of function, and definite, often severe, inflammation; it is therefore not surprising to find that in some individuals, especially predisposed, the gonorrhœal virus should cause an after-poison capable of producing the terrible effect which was seen in this case. Let me remind you of a fact, which I may have mentioned before and may have to mention again, because it is one of the cardinal facts of medical knowledge, throwing light on many problems, and, among others, on the influence of predispositions. It is the fact ascertained by the admirable researches of Dr. Sidney Martin on the process by which diphtheritic paralysis is produced. He seems to have conclusively proved this,— that this paralysis is due to a poison of organic, chemical nature; that the poison is not produced directly by the diphtheritic organisms, but that these generate a particular chemical substance of the nature of a ferment, which acts upon albuminous materials ("albuminoses") in the body, especially in the spleen, and converts them into a poison which has this intense action on certain nerve structures. It is probable that the mechanism is similar by which many other specific organisms give rise to toxic agents. You can readily understand that if a person has previously been in bad health, as this lad had been, there should be some slight defect in the chemical constitution of such substances as the albuminoses of the spleen, etc., and a very slight defect may render the product of such a ferment different from that which would result in perfect previous health. A very slight difference in constitution, a little more or less of one element, may change a toxic substance into a harmless one, and *vice versa*. Thus we can understand how it is that the same causal influence shall have but little effect in one individual, and shall have a profound effect in another, especially in another who has been in conspicuously bad health. We can understand, for instance, that the organisms of gonorrhœa shall, in one person, give rise to a toxic chemical material which acts on the joints, and causes the arthritis of "gonorrhœal rheumatism," while, in another, the poison produced shall act on the spinal cord and set up acute myelitis, as in this case.

The lad was in a state in which death seemed inevitable. In "acute ascending paralysis" (Landry's paralysis), as I have said, we have only an arrest of function by the blood state; in inflammation we have also a process of organic change, in the vessels and the tissues, which tends to progress even apart from its cause. The mere arrest of function, once stopped, may not increase, and may be followed by steady recovery; but the effects of inflammatory damage must persist, and the process itself may continue for a time after the blood state which has caused it ceases to act.

Hence, I fear that the chances for the poor fellow's life are very small. I doubt, indeed, whether he is alive now.

Although the expectation of saving life may be absent, and even the hope of doing so can find no room in reason, we are always under the compulsion of striving to achieve that which we may not be able to think possible. Our knowledge is never certain—even our possibilities may be incorrect, and so our hopelessness may be wrong. We must act in spite of our anticipations. Therefore, it will be right for you to ask, what did I suggest should be done for him?

As far as I have been able to see, there are only two agents which have any considerable influence upon toxic blood states of this kind. One is mercury. I believe the influence of mercury in syphilis is only one instance of its power over organisms. It is the most striking and perhaps the only one surely known to us, but it is probably not the only one, and is not likely to be the only one. Some years ago, a discovery was made in some researches at the Brown Institution which has been almost unnoticed since,—that the fatal effects of a certain inoculation (I think with the poison of anthrax) could be prevented by mercury. Its influence in other diseases of like nature is supported by many facts. That on inflammation, which few can doubt may be connected with its effect on blood states. So I advised that mercury should be rubbed in—a drachm of mercurial ointment every four hours. In such a case, however, you must "hit right and left;" there is not an hour to lose; you must do anything and everything you can, always provided you do no harm. In septicæmia, such septicæmia as occurs after childbirth, for instance, the agent which has seemed to me effective, and which, as far as I could judge, has certainly saved life, is perchloride of iron in full doses. So I advised that every three hours 20 minims of the tincture of perchloride of iron should be given. The only other opportunity for treatment, beyond ordinary stimulation, was this. He might die from failure of the functions of the medulla; they might fail from the influence of disturbance of other related parts, or from an influence which was not an actual inflammation; it was just possible that if they could be kept active he might tide over a period of danger. The measure that has seemed to me to have most influence in stimulating and keeping up the functions of the medulla, is the hypodermic injection of a small quantity of strychnia, one-eightieth of a grain, together with a very minute, stimulant dose of morphia, about one-thirty-sixth of a grain, repeated every two hours if necessary. Of course, if the failure of function were due to invasion by inflammation, this could have no effect.

But there was no reasonable ground for hope that the treatment would be effective. The cause is more likely to be a chemical substance than organisms, and we have even less power over the former than we have over the living agents of disease. Had the symptoms been due merely to the restraining influence on function exerted by the blood state, we could have felt a slender hope that its maximum might have been reached. But inflammation, to whatever due, is a process with its own effects—proportioned to its initial energy, but locally independent of its origin—certain, at such a stage, to cause more destruction. Already at the limit of the functions essential for life, it could scarcely stay there; the passage of that limit seemed inevitable, and with it the thread of life would break. Next week I shall doubtless be able to tell you.

[In his lecture the following week Dr. Gowers mentioned the sequel. The boy lived only six hours. A post-mortem examination, made by Dr. Arkle, revealed characteristic signs of myelitis, most intense at the place at which the inflammation had been supposed to begin—the lower part of the dorsal region, just above the lumbar enlargement. There, indeed, it had caused the front of the cord to be bulged forward in an unusual manner by the swelling and softening of its substance. There was also some softening in the mid-dorsal and lower cervical region, where the grey substance was hemorrhagic, almost diffluent. Culture-investigation for organisms has been commenced, and the pathological results of these and of the microscopical examination of the cord, will be published by Dr. Arkle.]

LECTURE IX.

LOCOMOTOR ATAXY.

I.

Gentlemen:—I desire to-day to direct your attention to the disease known as locomotor ataxy. It is a subject with which you are all probably more or less familiar, but perhaps not so familiar as to make it superfluous for you to consider afresh some of its features and some of the lessons they convey. The subject is so large that it will be impracticable to bring into the compass of a single lecture all the points to which I desire to direct your attention, and, therefore, I think it will be wiser for us to devote to it two of our weekly meetings. The subject is instructive on many accounts. The study of this disease supplies a useful example of pathological reasoning, and of the mode by which we discern the mechanism by which symptoms are produced. It is instructive also on account of the wide relation of its facts: those which you learn in considering it have an application to many other definite diseases, and to many other general morbid processes. It is also instructive because it is a typical example of the class to which it belongs,—typical not only in its nature, but typical also in its treatment. If you learn these thoroughly you will have learned also much relating to other diseases.

I propose to consider to-day the leading symptom of the disease, the ataxy, in reference to its mechanism—to consider it, not in confusing detail, but with perhaps superfluous deliberation. It is an advantage in clinical teaching,

Delivered June 14, 1893. *Clinical Journal*, September 27, 1893.

that information can be given in less compressed form than is possible in systematic lectures. In the latter, so many facts have to be described and discussed in such rapid succession, that, in the mind of the student, they are apt to jostle one another out of place, so that, at the last, perhaps only a very few facts remain prostrate in his mind. I propose to avail myself of this freedom, and to strive to fix in your mind only the leading facts relating to the subject.

There are few diseases of which the name has given more trouble than that of locomotor ataxy. This name was assigned to the malady by Duchenne, with the additional epithet "progressive," an epithet which soon came to be dropped on account of its clumsiness, and the dropping has since been confirmed by the discovery that the adjective was inaccurate. The disease in a large number of cases is not progressive. The ability to recognise the early stage of the disease has revealed to us the fact that at least half the cases, I think more than half, are not, or need not be, progressive. But the term "*locomotor* ataxy" is itself somewhat redundant, because the disorder is one of movement in general, and not merely of that which causes change of place. So "motor ataxy" would be as accurate as "locomotor ataxy." But the term "ataxy," as a simple term, is only applied to disorders of motion, so even the epithet "motor" has been largely discarded, and the term "ataxy" has been used alone. And yet it has been found that even this is inaccurate, because the disease may exist in its slighter form without even a trace of inco-ordination.

Hence the example has been widely followed, which was set in Germany, of using a still more general name, a term which had been before applied to various affections of the spinal cord, the term "wasting of the back," "tabes dorsalis." But, again, with the strange swing of the pendulum which seems inevitable in the progressive alteration of idea inseparable from the advance of scientific knowledge, we have found that the disease may exist when there is no affection of the spinal cord, and so it has been necessary to supplement the term "tabes dorsalis" with the term "peripheral neuro-tabes." I shall explain the necessity later on. It is rather surprising that in this difficulty more writers have not been disposed to fall back upon the name which has sometimes been applied to this disease, that of *Duchenne's disease.* The labours of that most distinguished and most energetic clinical investigator of diseases of the nervous system have been recognised by the ascription to him of the ownership of three maladies, —two diseases and one paralysis,—pseudo-hypertrophic paralysis, and locomotor ataxy are Duchenne's "diseases," and labio-glosso-laryngeal palsy is Duchenne's "paralysis." But this method of nomenclature has its disadvantages. It involves a grave burden upon the memory of the student, especially when it is applied to diseases and symptoms.

It has not only inconveniences; occasionally, in the admiration of one man, an injustice is done to others. I yield to no one in my admiration for Duchenne. Duchenne's description of the disease first called general attention to it, and the name he gave it was adopted. But what constitutes the *discovery* of a disease? If to distinguish the symptoms from those of other similar maladies, to affirm from physiological considerations that a definite structure in the nervous system is diseased in such cases, and that this disease is the cause of the symptoms, and then, in the body of a patient, to find and demonstrate that very structure to be diseased,—if this is not the discovery of the disease, I would ask you, gentlemen, what is. And that was done for locomotor ataxy by an English physician at King's College Hospital, Dr. Todd, and done nearly 15 years before Duchenne's description of the disease was published, and probably many years before he had begun to suspect its special character. If locomotor ataxy is to be called by the name of any physician it ought to be called "Todd's disease."

Let me now direct your attention to the general features of the cases that I have to show you, and then proceed to consider the leading symptom of the disease, its origin, and nature; next week we will consider in detail some of the other symptoms, and some of the general practical questions connected with the affection.

The first patient is a man aged 41. He can tell us of no apparent cause for the disease. But most of you know that there is one antecedent that must always be searched for—syphilis. His preceding history, however, affords no evidence of this disease. But those of you who were here a few weeks ago will remember that I insisted on the fact that the absence of a history of syphilis does not enable us to exclude the disease unless we can exclude the conditions of its causation. I need not repeat the considerations which compel this conclusion. We are not able to exclude it here. Three years ago he began to suffer from tightness and numbness around the chest across the front, with some pain there. These symptoms in the chest were followed by numbness in the feet, a year and a half ago, and then by shooting pains in the thighs, extending to the calves, heels, and arms, and accompanied by some unsteadiness in walking in the dark. He says also that some weakness of the legs came on; but patients with this disease very often describe weakness of the legs when there is none. They mistake the difficulty of co-ordination for deficiency of power.

Tardiness of micturition also developed, a symptom which is sometimes the earliest to attract the attention of the patient. Remember that many sufferers from diseases of the spinal cord first seek medical advice on account of difficulty of micturition; unless we are aware of the fact, they may be treated for a long time, and treated unsuccessfully, because the real cause is not suspected.

There was also transient double vision,—another early symptom of the disease, the relations of which we shall have to consider next week in greater detail.

He has now, as you see, slight inco-ordination of the left hand and of the legs, greater in the left leg than in the right. When told to touch his nose with his forefinger, he succeeds fairly well. With the left hand he can unbutton and button his waistcoat. If made to hold both his hands out, when his eyes are shut, there is very little spontaneous movement to be detected. When made to walk without the aid of his stick his gait is a little uncertain. When he puts his feet close together and shuts his eyes, you observe that he has difficulty in maintaining his balance. When asked to put his feet close together with his eyes open, he is just able to stand upright, not so steadily as is natural, and if then he raises his head up the tendency to sway is distinctly greater. In this position the base of support of the body is reduced in area, the muscular contractions have to be more accurately adjusted to maintain equilibrium, and the defect in the power of adjustment reveals itself. When, after placing my hand in a certain position, I ask him, with shut eyes, to touch it with his left foot, very little inco-ordination is revealed. His defect shows itself chiefly in walking and in standing. There is no knee-jerk in either leg. A little impairment of sensibility to touch can be found, chiefly on the outer surface of the legs. Reflex action from the skin is defective, and the reaction of the pupils to light is slight.

In the second patient, the gait is much more defective than in the first. His toes are raised too high as he walks. When asked to stand still with his feet wide apart he can do so, but with the feet close together he has great difficulty in standing. He separates his feet to increase the base of support; but even then the muscles are in constant action, so that the toes are every moment raised, because the over-action of one set of muscles is so great that it has to be compensated for by the contraction of another set, and this is also too great. Withdraw his visual guidance and he is quite unable to stand. After observing the position of my hand he cannot touch it with his left foot while his eyes are shut. His foot goes too much to the left. With the right foot he does better, but this also goes too high. He succeeds in touching his nose with the tip of his forefinger while his eyes are shut. When made to hold out both hands with his fingers extended, involuntary movements are seen, the action of the muscles which should keep the limbs at rest being disordered, like that of the muscles during movement. In this patient also there is no knee-jerk in either leg.

The symptoms in this case are of six years' duration; and the unsteadiness has been considerable, and accompanied by shooting pains for five years. For three years he has had symptoms of unusual character in certain cranial nerves which also we shall consider next week. The muscles of mastication are affected. He has difficulty in approximating his lips and likewise some paralysis of the ocular muscles. In addition, he has deficient reflex action in the pupils, impairment of sensation in the limbs, and also in the face.

In the other patient we could only recognise syphilis as a possible antecedent; in this, it is certain. We have distinct evidence of its occurrence. Twelve years ago he had the primary lesion, followed by sore throat.

These, then, gentlemen, are the chief features of the disease. We will now consider especially the characters of the leading symptom, that round which the others are grouped, as it were about a centre, and which, although not invariable, is yet the dominant element in the developed disease. In all cases, even when this symptom does not exist, conditions can be traced which are capable, ultimately, of producing it.

The *inco-ordination* of locomotor ataxy is a symptom which, as its name expresses, is in aspect a motor symptom. But it is not a motor symptom in its nature. It is not due to a primary derangement of motor processes and motor structures. In a typical case, if you test the strength of the limb in which there is the greatest disorder of movement, you find no deficiency of power, and if you examine it you find no trace of muscular wasting. Moreover, if you scrutinise with the microscope the motor structures, the cortex of the brain, the fibres that pass down the pyramidal tracts, the cells of the anterior cornua of the cord, the fibres that pass through the anterior nerve roots to the nerves, and even motor twigs in the muscles, you find no trace of disease. The motor structures are intact. But you find it very different when you examine the sensory structures. In all cases in which the spinal cord is affected—and they constitute the great majority—we find disease in the *sensory* part of the cord. Wherever there is inco-ordination there is sclerosis in the posterior columns at the place at which the posterior nerve roots from the seat of the inco-ordination pass into the cord and also degeneration of the posterior nerve roots. The existence of this disease where the posterior nerve roots enter the cord and of degeneration in the roots themselves, suggests that the essential lesion underlying the symptoms is an interruption in the conducting function of the posterior nerve roots. And that conclusion is remarkably confirmed by some rare cases in which the posterior nerve roots have been damaged by other causes. In one case especially (recorded by Dr. Hughes Bennett) in which the patient had multiple tumours upon many of the posterior roots, there were the typical symptoms of locomotor ataxy. This case is of great importance; it establishes the fact that the symptoms of the malady we are considering may depend on a simple lesion. It affords us one of the firm facts of pathology by which we may ascertain the meaning of other complex phenomena.

But the disease of the root-fibres is not the only lesion in tabes. All the way up the cord, above the place at which the root-fibres entering the cord are conspicuously diseased, we find degeneration of the posterior median columns—of that part of these columns that is adjacent to the posterior median septum,—the so-called "column of Goll." At the lumbar region, where, as a rule, the root-fibres are affected, this degeneration of the posterior median columns widens out, so as to blend with that at the entrance of the nerve roots. That is because the fibres which ascend in the median columns reach these from the roots by passing through the external parts of the posterior columns. It is because the median parts are composed of root-fibres that pass to them without interruption, and ascend in them without decussation. Hence disease of the lumbar roots necessarily entails the degeneration of the median columns, and its extension outwards to the root region of the affected fibres. The fibres degenerate upwards, and do so in the same way if the lumbar roots are destroyed by any other morbid process. It is thus certain some root-fibres ascend the cord without interruption, and that these fibres are diseased in tabes. That is the first great pathological fact for you to remember.

Now, leaving this point for the present, what evidence can we discern as to the way in which disease of the nerve roots causes the symptoms of this malady? Certain clinical facts, when taken together, have a very clear pathological significance. First, the leading symptom, ataxy, exists, in its typical form, and even in high degree, in many cases in which there is no impairment of any form of cutaneous sensation. Further, in agreement with this, we have the fact that when there is impairment of cutaneous sensation there is no proportion between it and the amount of inco-ordination. It is, therefore, certain that the disease of the posterior nerve roots does not cause ataxy by interrupting the impressions conveyed from the skin. Yet it is also certain that the symptom is due to the interruption of some impulses which these roots convey.

What other afferent impulses pass to the cord by these roots? They are the impulses from the deeper structures, from the muscles especially, also from the fibrous structures, from the fasciæ, perhaps from the tendons, and certainly from the joints. Remember that we must not confound afferent impulses and sensory impressions. It is probable that the afferent nerves are continually being traversed by impulses which do not act directly on consciousness, and do not, when in normal degree, produce a true sensation. We are unconscious of any sensations in our muscles at the present moment. But a vigorous compression of any muscle will at once convince you that afferent impulses can be readily generated in it; so also will an attack of nocturnal cramp, and so also will the process of going upstairs on the following morning. In cramp of the calf, the afferent nerve-fibres within the muscles seem to be stimulated by their compression in consequence of the narrowing and widening of the strongly contracted muscular fibres, and when the nerves are rendered hyperæsthetic by that disturbance, they are stimulated in a very marked degree by the tension to which they are exposed when the muscle is extended in mounting stairs. . . . So that we have evidence that afferent impulses are generated in these fibres by compression during contraction, and by tension during the passive elongation which the muscles endure on every movement. We can understand these facts by the mode in which these afferent fibres arise in the muscles. According to the investigations of Tschiriew, they commence in the interstitial tissue between the fibres, and must thus be equally exposed to the mechanical effect of compression or of tension. Thus, by the exclusion of cutaneous impulses we are driven to refer the origin of ataxy to the interruption of these deeper impulses. We can perceive more of the features of the impulses from the muscles, than of those from other deep structures, and the large extent of the muscular tissue makes its probable that they constitute the chief part of the deeper impulses. Moreover, the phenomenon of muscular co-ordination suggests that they, rather than the impulses from the joints, etc., take the chief parts in its regulation. This is confirmed by the constancy with which the knee-jerk is lost in locomotor ataxy. Whatever is the precise mechanism of the knee-jerk, it certainly depends upon a reflex process in which the afferent impressions proceed either from fibrous tissues or from muscles, and probably chiefly or exclusively from the muscles. But the constancy of that loss confirms the conclusion that it is by the interruption of these afferent muscular impulses that the ataxy is produced.

What is the significance of the degeneration in the posterior median column? We find it whenever the posterior nerve roots are destroyed. Moreover, if the nerve roots are destroyed on one side only, the degeneration of the posterior median column is on that side; this and the facts already mentioned show, as I have stated, that nerve-fibres pass from the posterior root, through the posterior column, into the posterior median column on their own side, *and do not cross.* And as this degeneration of the posterior median column is constant in locomotor ataxy, we cannot doubt that the fibres which constitute it are some of those the disease of which gives rise to the symptom in question. And that is confirmed by two other very interesting facts. If that posterior median column is divided on one side in a unilateral lesion of the cord, the effect is to cause incoordination in the muscles of the same side; whereas it is found that an extensive unilateral lesion causes loss of pain and sensibility on the opposite side. Further, in advanced tabes we commonly have a very marked loss of the muscular sensibility: the muscles may even be insensible to pressure or extension, thus confirming the conclusion that there is disease of their nerves, and supporting the opinion that this is the lesion which causes the characteristic ataxy of developed tabes.

Where do these posterior median fibres go? You must often have been struck by the similarity between the mode of standing and walking in some cases of ataxy, in which the power of maintaining equilibrium is especially deranged, and that of a patient with cerebellar disease. These posterior median fibres go up to the posterior median nucleus of the medulla; and it has been proved that there are many other fibres proceeding from the cells of that nucleus to the middle lobe of the cerebellum, the part of the cerebellum that is alone concerned in coordination. The probability is thus very great that these fibres conduct impulses which reach the middle lobe of the cerebellum and guide it in its co-ordinating function. So that we seem to perceive two processes in the mechanism of co-ordination: (1) Impulses from the muscles go to the cord; some of these must enter the grey matter to subserve the reflex action on which the knee-jerk depends, and they probably regulate a series of subsidiary co-ordinating mechanisms. When you stand, you stand by a voluntary impulse, but it is probable that the voluntary impulse acts through a mechanism in the cord which determines the arrangement of the movements for standing. The action of this mechanism is conspicuously seen in cases in which disease at the middle of the cord cuts off the lower part. Then, as you well know, spastic paraplegia is produced, and the extensor spasm perfectly reproduces the condition of the limb in standing. That extensor spasm, which is attended by a gradually increased knee-jerk, is apparently dependent upon arrangements which are acted on by these impulses from the muscles. (2) Impulses pass up the cord, and probably reach the cerebellum, and through the cerebellum they may act on the motor cerebral cortex, and influence the relative energy of its cells. Thus we have this double reflex process of co-ordination from the muscular impulses, one by the cord, the other through the cerebellum. When a sensation of pain is produced in a muscle by a strong impression on its nerves, it is probably through those that terminate in the cord; but we have no direct proof of this, and it has no special concern for us in our present problem. We perceive, then, that muscular inco-ordination is due to loss of impulses from the deeper structures, chiefly from the muscles. If there is a total loss of all the impulses, the muscle reflex is abolished, the knee-jerk is lost, and we have extreme ataxy. If there is only a loss of the impulses that pass up to the cerebellum, when there is disease of the posterior column above the lumbar enlargement (as in ataxic paraplegia), the knee-jerk is not lost, and we have only a moderate degree of inco-ordination, manifested especially in the defective maintenance of equilibrium, and almost identical with that of cerebellar disease.

I mentioned, just now, that the spinal cord is not invariably affected in tabes. Exceptions to the common rule are extremely rare, but their occurrence is well established. When the cord is free from disease the peripheral nerves have been found to be degenerated. They suffer, indeed, in a large proportion of the cases in which there is degeneration of the root-fibres; the peripheral endings are altered in chief degree. In the cases in which the cord is unchanged this lesion of the nerves, usually concurrent, is isolated, and the sole cause of the symptoms. These do not differ from the symptoms of cases with the usual lesion. Hence, we are justified in concluding that the nerve-fibres, the disease of which gives rise to the characteristic symptom, ataxy, are the same in both forms. The nerves which are affected first and most in these cases, as in the others, are the nerves of the muscles. So that, whether we have disease of these muscle nerve-fibres in the posterior nerve roots and posterior columns of the cord, or disease in their peripheral extremities, we have precisely the same condition; if the mechanism is the interruption of the conducting fibres, the effect must be the same, wherever in their course the lesion is situated. Hence, in "peripheral neuro-tabes," as the rare form is termed,—locomotor ataxy without disease of the cord,—we have a condition precisely similar to that which depends on the spinal lesion. Hence, also, there is ataxy similar in its features in the other forms of degeneration of the peripheral nerves, such as are produced by Alcohol and by Arsenic. It may even be impossible, from the symptoms themselves, as far as the ataxy is concerned, to distinguish a case of true locomotor ataxy from one of what is called "alcoholic pseudo-tabes."

I hoped to have gone over to-day other points regarding the nature of the lesion, but it will be better to postpone these until next week. I must be content to-day if I have succeeded in giving you a clear conception of the mechanism of the special symptom of the disease, or at least of the more salient elements of the evidence regarding its nature. I would only point out to you, in conclusion, how interesting is the enlargement of our ideas regarding this disease which has come from modern research. There is much that we cannot understand concerning the precise mechanism of co-ordination and of its defect; nevertheless, we have a glimpse into the process, though we cannot surely perceive all its outlines. It is certain that the afferent nerves of muscle are stimulated by the changes which they must endure in the two states of the muscle; the muscular contraction when the muscle is active, and the tension it undergoes when its opponent is acting, alike generate afferent impulses. These we do not ordinarily perceive, except in the vaguest manner. They are doubtless concerned in the production of our conceptions of posture and movement, but we only perceive them as sensations, when from some cause they become excessive. But they must be constant, in unperceived degree, influencing and guiding the central mechanism.

Consider the subject over at your leisure, and I think it will become clear to you, and that you will find assistance from it in understanding many processes and many problems to which I have not ventured to refer.

LECTURE X.

LOCOMOTOR ATAXY.

II.

You may remember, gentlemen, that last week we studied at considerable length the chief symptom of locomotor ataxy and its mechanism. To-day we will briefly review the other symptoms of the disease. They are, however, so numerous that I shall have to content myself with a mere glance at those that are of chief importance and at their relations, considering in more detail some that are illustrated by the cases we examined, or that are of most significance, or with which you are not likely to have become familiar.

The morbid process which, as you saw in the microscopical sections, is indicated by an overgrowth of the connective tissue, is essentially a degeneration of the nerve elements, beginning in them. Such degeneration has many causes. It seems sometimes to be the result of a mere inherent defect of vitality; sometimes it is a local manifestation of senility; sometimes it is due to the influence of general conditions, depressing emotion, and like influences; but in the vast majority of cases it is the result of the action on the nerve elements of a toxic agent. We see it as the effect of chemical agents of inorganic nature, and also as the effect of certain organic chemical agents which are produced by minute organisms—such, for instance, as alcohol, which is due to the growth of minute organisms outside the body, and materials of unknown composition produced by the

Delivered June 21, 1893. *Clinical Journal*, October, 1893.

minute organisms which constitute the virus of some acute diseases. Such organisms cause their own direct effects, but they also leave behind them in many cases such a substance as I have referred to, which acts upon nerve elements as inorganic poisons may do, and causes degeneration. And this, which has been proved to be true of the virus of some acute specific diseases, seems also to be true of the chronic specific disease which is so frequent an antecedent of ataxy, syphilis. This was, you will remember, certainly traceable in one, and possibly in the other, of the patients whom we saw last week, and whom we will presently again examine. As a fact, this antecedent is to be ascertained in more than two-thirds of the cases; and it has probably preceded tabes in at least four-fifths. Such a relation means a causal influence in the majority, for mere coincidence will explain only a small proportion. At first the statement of these facts was received with general scepticism. But we have to be careful, in science, how we deny. Some denied the possibility of the causal relation, but they have had to reconsider their denial; fresh observation of facts yielded evidence of overwhelming cogency. One reason why the relation was doubted was because it was found that anti-syphilitic remedies did no good to the disease; and the fact is unquestionable. But the mystery is explained when we perceive that only on the lesions which are the direct result of the syphilitic organisms does anti-syphilitic treatment exert its influence; on those lesions which are the result of the toxic material produced by the organisms, the anti-syphilitic treatment has no influence. This material has not been isolated, but the analogous material has been isolated in some acute diseases dependent upon organisms; and so close an analogy can be traced between the whole series of such diseases which have after effects, that we may feel reasonably sure that it is in this way that syphilis produces the disorder we are now considering. It is not possible for me now to describe to you the evidence, but it is, in my opinion, convincing. In these facts we have the secret of the inutility of anti-syphilitic treatment in cases of locomotor ataxy, in which the disease is, nevertheless, the sequel of syphilis.

You will remember that we traced the chief symptom of tabes, the ataxy, simply to the interruption of the muscle nerves and the arrest of the afferent impulses which guide alike the spinal and the cerebral motor centres. You will also remember that I pointed out to you that, in this mechanism, it makes no difference in what part of the nerve fibres the interruption takes place, and that when the ends of the nerves alone are diseased, the effect is the same as when the same fibres are diseased in the nerve roots or within the cord.

A curious historical fact may be mentioned in this connection. In what may be termed "medical surgery," from time to time, some operative procedure becomes fashionable; it is adopted on all sides, partly on the strength of an initial laudation, and partly because it is something to "do" where little had been effected. After a time, if it possesses no real value, it sinks into desuetude. Of late we have had an epidemic of "suspension" for this malady. A little before, it was nerve stretching which was said to do remarkable good in cases of tabes. The curious fact to which I allude is this, that the first patient who seemed to derive benefit from nerve stretching was about to undergo the procedure a second time, but died under the anæsthetic before the operation could be performed, and so afforded pathologists an opportunity of correlating the operation and the morbid process. The operation, unfortunately, was only partially utilised. It was found that there was no discernible disease of the spinal cord or of the nerve roots. The peripheral nerves were not, in that case, examined; they have been examined in others and found diseased in so many, that there can be little doubt that the case I have just referred to was one in which the degeneration was at the periphery of the nerves. I mention this fact to impress on your minds the occurrence of this peripheral change. It has been found to be exceeding common, if not constant, in tabes, although it is very rarely confined to this position. If it is the only lesion, there is no disease in the posterior roots. The disease above the ganglia seems to be an independent, coincident, degeneration in the same fibres.

The disease is not limited, however, to the afferent muscle nerves (to which I especially directed your attention last week), although their affection seems to be the primary and essential element. The nerves from the skin suffer in a large proportion of the cases, and, in consequence, loss of sensation to touch and to pain and to heat (usually to some, occasionally to all) is very common, and is distributed according to the position of the nerve lesions. The anæsthesia, like the ataxy, may be due to disease either of the nerve roots in the cord or in the periphery. It is found sometimes in the arms, and there is, occasionally, a similar loss in the cutaneous sensibility of the head. We may have anæsthesia in the region of the fifth nerve; and degeneration of the optic nerve, as you know, is not rare. The latter is a typical nerve degeneration, involving the peripheral part of the nerve, and is a visible type, in a nerve of special sense, of that which occurs in the sensory nerves of the limbs.

The process of degeneration is one of molecular change. But the function of the sensory nerves is carried on by molecular change. It is thus that they perform their marvellous function of transforming, for instance, the mechanical force of motion, in a touch, into nerve energy, or the thermic force, which is another form of motion, into nerve energy. And the molecular changes of degeneration seem to give rise to nerve impulses similar to those which are produced under normal circumstances,—similar, I mean, as regards their general character, not as regards their special occurrence. Hence we have the symptom, so familiar to you, of the pains of tabes. Most common are the peculiar, sudden, sharp, "lightning" pains; but duller pains are very common; and they are often curiously "rheumatic" in character, so that they are almost always at first mistaken for rheumatism. So, indeed, with less excuse, are the sharp momentary pains. This mistake is facilitated by the fact that change in weather, especially change to cold, has a remarkable effect in increasing these pains, just as it does the pains of true rheumatism, and many patients have been treated for several years as the subjects of rheumatism or of gout, whose pains were the result of this malady. These are felt most often in the legs, but also sometimes in the trunk and sometimes in the arms, and occasionally with great severity even in the head. In the head and trunk they are apt to be mistaken for simple neuralgia, and even thought to be the indication of encephalic or abdominal disease.

Thus, these being the essential symptoms of the disease, it is by its nature a sensory malady; it is an affection of the sensory structures. Its dominant feature, the inco-ordination, is, as I told you, motor only in aspect. The cardinal symptoms of the disease are the result of changes in afferent paths. They are, however, often associated with subordinate symptoms which are actually motor. All toxic agents which may act, primarily and chiefly, on sensory structures, can also act, to some extent, on motor structures, and some of the agents often act on these in a preponderant degree.

Among the motor structures which are most often impaired in tabes are the nervous structures for the bladder, sometimes those of the sphincter, but more commonly those of the wall of the bladder. The wall of the bowel may suffer also and give rise to constipation; this is inconvenient, but the weakening of the wall of the bladder involves definite danger. In consequence, the bladder is not emptied; there is residual urine, and that which descends from the kidneys overfills the bladder; the constant overfilling necessarily weakens still more the wall, because when overdistended its muscular power is relatively more incompetent, and it is less and less able efficiently to contract. Chemical changes occur in the residual urine, and cystitis is excited, with a secretion that increases the alteration. The increased pressure in the bladder, thus gradually augmented, acts back on the kidneys, interferes with their function, and ultimately induces disease in them of the nature familiar to you as surgical kidney. The blood, from the first, is not properly relieved of effete material, which seems to become, at last, actually toxic, and a state of imminent peril to life is often thus induced. Of all this chain of effects the patient may be unaware. The overfilled bladder indeed often overflows, and incontinence reveals the condition.

But overflow does not always occur, or occurs so seldom as to attract little notice, and lowered sensitiveness may aid in preventing any consciousness of what is wrong. Remember that you must never take a patient's statement as true that he empties the bladder. Patients are constantly under the impression that the bladder is emptied, when you may easily ascertain by percussion over the pubes that a considerable amount of urine remains in the bladder after micturition. This evil may reach a high degree in a patient who is in the early stage of the disease. It is the chief way in which locomotor ataxy causes death; and death in this way can always be prevented. Whenever you suspect residual urine, ascertain whether your suspicion is correct by passing a catheter. Remember that it is absolutely essential for the bladder to be perfectly emptied at least once a day; and if there is any decomposition of the urine or distinct muco-pus in it, the bladder must be daily washed out with a disinfectant. The effects of the retention on the kidneys and their function when the patient has been brought to the edge of disaster, are sometimes precipitated by the reflex effect of the passage of a catheter; then the patient is said to have "catheter fever," but the condition would have developed spontaneously in a few days, and is due essentially to want of the previous use of the instrument. Only two days ago I saw a gentleman whom I saw also a year ago in the early stage of this disorder. At that time he was having mysterious attacks every few weeks, in which considerable pyrexia lasted for several days. The doctor who was attending him had been greatly puzzled by these, and suggested various theories to explain them; but I found the patient was not emptying his bladder. Such attacks of pyrexia in that association are generally due to the interference with the action of the kidneys and the effect on the blood. The patient commenced the use of the catheter, had one more slight attack of pyrexia, had since used the catheter every day for a whole year, and had not had another attack.

I ought to add, to avoid misconception, that incontinence is not always due to overdistension, any more than overdistension always causes incontinence. In many diseases of the spinal cord you have paralysis of the sphincter, urine is always escaping, and the bladder is always almost empty. In some others, the bladder is periodically emptied perfectly, by a pure, involuntary, reflex, or automatic process. But in locomotor ataxy incontinence is generally due to overfilling.

Of other motor symptoms, the next in importance are those with which we meet in the eye, of which the most common is the loss of the reflex action of the iris to light.

This seems to be due to changes in the centre for this movement in the nucleus of the third nerve. Instead of this loss, or with it, there may be loss of accommodation, or loss of the action of the iris associated with accommodation. Frequently there is weakness of external ocular muscles, sometimes amounting to complete paralysis. Such weakness is usually at first transient, but after several successive attacks, which last longer and longer, the loss of power finally becomes persistent, and it may be present in muscles of both sides. All these ocular symptoms are probably due to central degeneration in the related nuclei.

There is one important fact connected with this loss of action of the pupil to light. It is exceedingly common in tabes; tabes generally has syphilis for its antecedent, and this is also true of the isolated loss of reflex action to light, apart from the other symptoms of tabes, or even from any symptoms of disease of the nervous system. Whenever you meet with it as a persistent, isolated symptom, you will always be justified in suspecting that the patient has had syphilis, and in most all cases you will find the suspicion verified. This indication is often of very great importance as putting you on the track of syphilis when you would not otherwise suspect it.

Other cranial nerves rarely suffer in tabes; but in one of the patients I show you, by the courtesy of my friend, Dr. J. Taylor, under whose care the man is, there is a remarkable affection of the motor cranial nerves; not only the eyeball muscles, but the muscles of mastication are weakened, while the face, palate, and larynx are also all more or less paralysed, one vocal cord being motionless.

Remember this case as an example of the wide extent of cranial nerve-palsy we sometimes meet with.

There may be some actual loss of power in the limbs,

but only in late cases of true tabes. It is due to the fact that, ultimately, even the spinal motor structures may occasionally suffer; but such weakness is not part of the malady in its common form.

Allied to the pains (which, I should have added, are often curiously paroxysmal) we have other paroxysmal symptoms which I need only mention—those which are called "visceral crises"—gastric pain and vomiting, laryngeal spasm, nephritic pains, rectal pains. These I can only mention to-day. But the affection of the sensory nerves has often another and very important consequence. The sensory fibres of the posterior roots are those which govern nutrition in all the structures except the muscles. We do not know how they govern nutrition; indeed we can only say that their own nutritional state seems to determine that of the tissues, so that the latter is disordered when the nerves are actively diseased. If they are the seat of acute inflammation, such as may be produced by passing a thread with some irritating material through the nerve, we have acute changes in the skin, but when there is merely chronic disease, slow, and without irritation, we have only a gradual thinning of the skin and wasting of the bones. For instance, such disease of the nerves of the arm gives rise, in the finger-tips, to thinning of the skin, and distinct narrowing of the ends of the phalangeal bones, and often also changes in the joints, by which adhesions form. When you have a peculiar irritation of the nerves, such as gives rise to the severe pains of tabes, changes may occur in the nutrition of the skin. The irritation of the nerves is shown by the intense hyperæsthesia that developes at any spot in the skin to which the pains are referred. The influence on nutrition was curiously shown by one patient who had such localized tabetic pains in various spots on the hairy scalp. After the pains had lasted for a few days every hair broke off near its root, and when the pains ceased, the growth of the hairs became natural. Among the most common trophic changes, however, are those of the bones and joints, when, instead of wasting, as they sometimes do, they enlarge, and new bone forms in an irregular manner. This may be due to an irritation, rather than to a mere defect of nerve influence. Its occurrence is an interesting illustration of the degree and extent in which the deeper nerves suffer in this disease.

One other trophic disturbance is both common and of great practical importance. It is the tendency to deep "perforating ulceration" on the toes, consequent on slight injury. It may reach the bone and lead to necrosis. These ulcers are often set up by the deep cutting of a corn. Always impress upon a tabetic patient that no corn should ever be cut; it may be softened by an alkali, and rubbed with sand-paper, but should never be cut.

The symptoms of tabes are thus, you see, numerous and varied; and according to their grouping they present very different aspects. If you remember the leading features you will seldom have any difficulty in arriving at a diagnosis. But there are certain special sources of error in diagnosis which it may be well briefly to point out to you. I cannot attempt even to glance at all the elements of the diagnostic problems presented by the disease, and must limit myself to a few which are either of special importance, are instructive, or are likely to escape your notice until the occasion arises for using the absent knowledge. In patients who are in the first stage of the disease, before ataxy has developed, and in many of whom it never does develop, uncertainty is sometimes caused by the limitation of the chief disease to the dorsal region. The pains are then felt in the trunk, and may be thought, and often are thought, to be due simply to intercostal neuralgia. But loss of sensation accompanies them; and if you examine the legs carefully you will always find the knee jerk absent, even when there is no anæsthesia of the limbs, and no pains are referred to them. Such a state has also been mistaken for growth in the spinal bones, but in this, the terrible pain is especially related to movement of the spine, and that relation is never present in tabes.

Other difficulties in the recognition of the malady may be due to the unusual character of the symptoms, or to the preponderance of special symptoms which may bring the patient under treatment. Many ataxics come to a doctor first simply because they cannot easily pass water. Whenever that is the case, without local cause for the difficulty, always suspect this or a similar condition, and test the knee-jerk. In other cases the gastric crises have been thought to be due simply to primary gastric disorders. These difficulties are chiefly due either to imperfect knowledge or to incomplete examination. But there are other and more excusable diagnostic difficulties which arise from the occasional difficulty of distinguishing tabes from allied diseases. There are other morbid states which present almost the same symptoms. Especially is this true of chronic multiple neuritis to which I have already referred. When due to alcohol or arsenic it may be manifested by pains similar to some of those of tabes, as well as by loss of cutaneous sensibility, and also by loss of the knee-jerk, and distinct, even conspicuous, inco-ordination. This is true, as I have also mentioned, of the ataxic form of diphtheritic palsy, although, in this, pains suggestive of tabes are usually absent. But in all these cases the chief indication depends upon the detection of the cause, which you will seldom have difficulty in discerning, if only your scrutiny is properly directed by a knowledge of the possibility. Often also your opinion will be supported by the loss of pupil-action to light, which is so common in tabes, but which is very rare in the maladies that may be mistaken for it.

Another class of diagnostic difficulties is due to the fact that there are cases which present all the symptoms of tabes, but others in addition. The most frequent and most important of these are cases of general paralysis of the insane. As a matter of fact, these patients *have* tabes; not only are its symptoms present, but its lesion exists; but they have also the cerebral degeneration which causes the special symptoms of general paralysis. It is not a question of distinction; it is a question of recognising the combination. As a rule, also, the common antecedent of pure tabes, syphilis, has also preceded the combined malady.

Thirdly, there are cases which present some symptoms of tabes, only others are absent and are replaced by symptoms of a different character. In the malady which we term ataxic paraplegia there is conspicuous unsteadiness of movement and in standing, but there is weakness as well; the knee-jerk is not lost; pains are slight; the pupil is not affected; and, as the weakness increases, there is an increasing tendency to extensor rigidity in the legs; and, with less power of movement, its disorder necessarily sinks into the background. This malady is due to a combination of sclerosis in the lateral columns of the cord, and in the posterior columns in the dorsal region, involving especially the ascending fibres of the posterior median columns, the fibres which, as we have seen, probably convey impulses to the cerebellum. It does not involve the fibres which convey impulses to the grey matter, at the level of origin, which subserve the knee jerk. So the knee-jerk is not lost; the irritation of the nerves, which causes the pain, and their interruption, which causes the anæsthesia and the extreme ataxy, are absent. Wanting these features of tabes, the cases present the loss of power that is absent in typical and early tabes, and a sufficient examination, with adequate knowledge, renders the nature of the malady clear enough.

In cerebellar tumour there is unsteadiness which reminds us of some cases of ataxy, and of the cases that I have just described. But there are not the other leg-symptoms of tabes, with one curious, but not frequent exception. Occasionally, the knee-jerk cannot be obtained in cases of cerebellar tumour. It is a curious fact, but we cannot yet give a satisfactory explanation of it. It is especially curious on account of the association with unsteadiness of movement in both diseases—an association which is probably not accidental, but has a significance at present obscure. In cerebellar tumour there is no anæsthesia in the limbs; very often, instead of an absent knee-jerk, this is increased, and there is even a foot-clonus from pressure on the pyramidal tracts; headache and vomiting are conspicuous features, and the latter does not occur in isolated "crises," as in tabes. Lastly, optic neuritis is extremely common, and atrophy, if found, is distinctly that which succeeds inflammation, and quite different from the atrophy of tabes.

Lastly, some hysterical girls present some unsteadiness in walking, and the question of this disease has arisen. But they have no other symptom of tabes. The ataxy is trifling and limited to the maintenance of equilibrium. It may not be easy for you to conceive that such patients should be thought to be suffering from tabes, but this error is sometimes rendered more easy by the fact that no knee-jerk is obtained. In many cases of hysteria, and, indeed, in many perfectly healthy persons, the knee-jerk seems to be absent; sometimes many attempts fail to obtain it. This is, however, simply due to the fact that the muscles are not relaxed, and a very slight voluntary contraction of the flexors prevents the jerk. I believe the knee-jerk is never absent unless there is structural disease in muscle, nerve, or cord, or (rarely) within the skull. Some fifteen years ago I published a number of cases in which it was absent under normal conditions. All I can say now is that I should like, very much indeed, to examine these patients again, because I have no doubt whatever that I should now succeed in eliciting the jerk in every one of them. So, when I am told that the phenomenon is absent in a healthy person, I venture to doubt the fact, and in doing so I only treat the observations of others as I treat my own past observations. If the subject shuts the eyes, hooks the flexed fingers of the hands together and pulls on them, the jerk generally becomes distinct, though before it may have seemed absent. In a few, it may be necessary to let the legs hang vertically over the edge of a table, and it is always well to place your fingers on the hamstring tendons to ascertain their degree of relaxation. Yet, with all these precautions, I sometimes meet with cases in which I cannot obtain it on repeated attempts, and yet, next day, it is produced in such typical form as to prove that its absence had been apparent only. Regard then with doubt even your own observations on cases in which there are no other symptoms usually associated with its loss.

What of the prognosis of this disease? First, remember that the malady involves very little danger to life, however extreme it may become, save by exhaustion due to laryngeal or gastric crises, or by the insidious development of some disease, such as pleurisy, which may not cause the warning pain that it produces with normal nerves. But one frequent and serious danger to life is directly related to the disease —that from the affection of the bladder, and impairment of the functions and structure of the kidneys, which I have already described. This danger may be prevented. On the other hand, unfortunately, actual recovery is even less likely than is death.

The prospect, however, is not so gloomy as this statement may suggest. I speak of " recovery " in the medical and accurate sense of the word. Short of the disappearance of all the symptoms of the disease, they may and often do lessen to a degree that is practically unimportant to the patient, and they still more often cease to increase, and, though continuing, are tolerable. In the majority of all cases, the disease, carefully treated, is not progressive. This is true at whatever stage it is met with. Only when the symptoms are rapidly or steadily increasing (especially if this is in spite of treatment) must you be apprehensive that the affection will go on to grave disability. On the other hand, in every stage you are not justified in holding out to the patient the anticipation of more than arrest and a slight degree of improvement. More may be achieved, but you cannot reckon on it. In the early stages, arrest is almost tantamount to recovery, so far as the patient's consciousness is concerned. You cannot say that the patient has actually recovered, because if you test the knee-jerk every year for twenty years, you will probably find it still absent. But, happily, the loss of the knee-jerk is not a symptom which enters into the patient's consciousness unless it is assisted there by medical help. The pains, it is true, are not likely to leave consciousness untroubled, but they often become trifling and can generally be relieved.

Remember, moreover, in connection with the pains, this very important fact, that tabetic pains, however severe, do not necessarily mean continued increase of the disorder. They seem, once excited, to go on as a neuralgia does—without the morbid process becoming greater. They are merely a persistent effect of the persistent alteration of molecular nutrition.

It is unfortunate that, in considering any disease, we are compelled to leave to the last the subject that is first in importance—its treatment. I have already told you that, in spite of its relation to syphilis, nothing is to be looked for from anti-syphilitic treatment. In very rare cases, of early stage and rapid cause, iodide of potassium has apparently done some good. In many of these its influence has been combined with that of other agents, and it sometimes does good in diseases that are not related to syphilis. Certainly your experience will soon demonstrate to you the accuracy of the general statement I have made.

But if nothing is practically to be looked for from anti-syphilitic treatment, much is to be dreaded from its energetic use. When a patient has passed the stage of active syphilis, and is suffering from the sequelæ due to its residual influence, such treatment, and especially mercurial treatment, seems to intensify the tendency of the nerve elements to degenerate. I have seen some disastrous cases, in which a grave affection of this character has immediately followed a thorough course of mercurial treatment, what they call abroad a " cure "—a term which bears to us more unconscious irony than it does to those who employ it. True, in many cases it may be wise to give the patient, for a few weeks, some iodide of potassium, because he is sure to be told that he ought to have had it, and it does not produce the harm of thorough mercurial treatment. It is often well to clear the therapeutic ground of any possible source of future, though baseless, feeling that anything has been neglected, provided the measure involves no harm to the patient. Moreover, that which is true of full doses of mercury is not true of small doses. There seems to be some truth in the assertion that very small doses of mercury have a tonic influence on patients who have had syphilis, but who have ceased to be actively syphilitic. Such doses may, therefore, often be wisely given in combination with other agents. But, in treatment by drugs, it is essential to realize the fact that the state and process you are striving to modify and arrest is essentially a degenerative process, a process of defective nutrition, tending to structural decay. As far as we can discern, it presents no difference from the similar condition that is due to other causes. Further, the medicinal agents that sometimes exert a distinct influence on the process, seem to do so irrespective of its causation. Among these agents the first place must be given to arsenic. Its influence is, in many cases, conspicuous to no one more than to the sufferer. It is very interesting to note that we may trace a reason for the effect in the fact that arsenic has been proved to influence the nutrition of the peripheral nerves, especially the afferent nerves of the skin, which conduct sensation, and which also, as we have seen, determine nutrition; probably it has a similar influence upon the deeper afferent nerves. It has to be continued, however, for two or three months, and then, after one or two months omission, it may, with advantage, be resumed. The form may be either the liq. arsenicalis, or liq. arsenici hydrochlorici, in a neutral or acid mixture, respectively, or the arseniate of soda in a pill. Phosphorus, which has so many points of similarity to arsenic, may be given alternately, and strychnia or nux vomica may be combined with either.

Next in value to arsenic I am inclined to place a drug which I have used in many cases—seldom without benefit—chloride of aluminium. I do not know whether it has been used by others in such cases. It seems to have a combined tonic and sedative influence on the structures affected. Only yesterday I had a letter from a doctor about a patient who said he had been taking chloride of aluminium some time ago with great benefit, but had lately been trying a number of things. The doctor informed me that with none of these did he get any improvment, rather, his walking and his pains became worse, but as soon as he resumed the chloride of aluminium he again began to improve. It seems to have an especial effect in lessening the tendency to the pains. The dose is two, three, or four grains twice or three times a day after meals. It is fairly soluble, and you may give it in combination with belladonna, Indian hemp, or nux vomica. Other tonics, as quinine, and sometimes iron, may be given with advantage in combination with arsenic or aluminium. Paroxysms of pain are most effectually relieved by antipyrin or acetanilide (once called antifebrin). When superficial they are usually lessened by the application of chloroform to the skin, and very certainly, for several hours, by the hypodermic injection of cocaine at the upper part of the area of the skin to which they are referred. It does not influence the deep pains.

No other local measures are of distinct use. Electricity is not generally of service. If, indeed, the muscles are energetically faradised, so as to stimulate and wake up, as it were, the afferent nerves, the patients seem afterward to walk a little steadier; but I have not been able to feel sure that permanent good has resulted from a course of this treatment.

These, then, gentlemen, are the chief points with regard to the disease, to which I desired to direct your attention. I must hope that the facts I have mentioned, whether any of them are new to you or not, may be, at any rate, fixed in your minds by the relations I have pointed out, and those which have a direct relation to practical medicine may prove of service.

We will now look again at the two patients whom we saw last week, and examine them in the light of what I have been saying to-day.

I here show you the patient with the remarkable affection of the cranial nerves—to an extent, indeed, that we very seldom see. He has, as you observed last week, considerable ataxy, loss of knee-jerk, loss of sensation; he has even some loss of sensation in the region of the fifth nerve. His pupils are large, and do not act to light or to accommodation. Accommodation is probably paralysed. There is weakness of the right and left external recti muscles. The eyes are raised fairly well. The left converges much better than the right, although the right moves well in the lateral movements,—an illustration of the distinctness of the centres for the movement of the internal rectus, with its fellow, in convergence, and that with the external rectus, in the lateral direction. There are conspicuous hollows in the position of the muscles of mastication, the temporals and masseters, and you can hardly feel any muscular contraction there. The motor branch of the fifth nerve on each side has evidently degenerated in extreme degree. The tongue is protruded well. He cannot bring his lips perfectly together, in consequence of weakness of the lower part of his face. He closes his eyes fairly well. He seems to close his glottis (as you observe by the explosive cough), although one of the vocal cords was found not long since to be paralysed.

In the second patient there is a remarkable defect of sensibility upon the chest. He is suffering from the form of tabes to which I have already referred, in which there is greater damage to the nerve roots in the dorsal than in the lumbar region, so that there is a wide band of anæsthesia around the thorax. He has no pain, only tightness round the chest. The irritation in the nerve roots has only caused this sensation of constriction. This sensation occurs very frequently when the root fibres are damaged and irritated by other processes. It is not a rare symptom, however, in locomotor ataxy, so that you must not imagine, because you meet with it, that there is necessarily any focal lesion of the cord.

Just below the nipple he ceases to feel a touch. Between the nipple and the umbilicus there is anæsthesia on the right side, but only in a much narrower band on the left. There is also a more extensive, wider affection of the nerves of touch than of the nerves of pain. In the first patient there was a loss of the plantar and cremasteric reflexes, in the second the abdominal reflex should be lost in a greater degree on the right side than on the left. (On examination this was found to be the case.) It is distinctly greater on the left side, on which the sensation of touch has suffered most impairment. The loss of reflex action from the skin is related in greater degree to the loss of tactile than to that of painful sensibility.

Here I must stop. Let me beg you, however, to go on. Supplement any new knowledge you may have gained to-day by reading a systematic account of the disease; strive to connect the facts you thus learn with those you have just observed; and, above all, let the knowledge give point and effective power to the observation of the cases that are sure, sooner or later, here or elsewhere, to come under your notice.

LECTURE XI.

THE FOOT CLONUS AND ITS MEANING.

Gentlemen:—The patient before you presents an unusual degree of increase in the symptoms associated with the name "tendon reflex," of which the knee-jerk and foot clonus are our special types. I have lately had occasion to draw your attention to the loss of these processes, and to the relation of the loss to the inco-ordination with which it is so often associated. The features of the excess of these phenomena, which I am going to show you, suggest that we may wisely consider their excess in relation to their nature. The process on which they depend has been, in my opinion, widely misunderstood, and a correct conception of its nature makes clear to us much that would otherwise be obscure. It is worth while, therefore, to consider its character, although in doing so we shall have to traverse ground familiar to many of you. I find, however, by experience, that the knowledge supposed to be familiar is often in a confused and shapeless state, and it is as useful to reconsider such a subject as it is to present the actual novelties that are welcomed with much more eagerness.

What is the knee-jerk? When the leg is in a certain posture, a tap upon the patellar tendon makes the muscles contract that end in it; and, when the foot is in a certain posture, a tap on the Achilles tendon makes its muscles contract. It was at once assumed, not unnaturally, that the tap stimulated the nerves of the tendon, and that the contraction of the muscles was a reflex action. At that time, however, the presence of nerves in the tendon had not been ascertained by the microscope; the name employed, however, involved their existence. A "tendon-reflex" action could not occur unless the tendons contain nerves. Hence, the point was at once investigated and the result was to prove that there are nerves in the tendon. But their investigations were wholly unnecessary to establish the fact. May I ask each one of you now present to compress suddenly your Achilles tendon between the thumb and finger? If the squeeze is energetic you cannot possibly need the microscope to prove the presence of sensory nerves in the structure. The discovery, therefore, that nerves could be seen in tendon was superfluous, but it supplied the needful evidence that there was a mechanism for a "tendon reflex," and so it seemed to confirm the view that the contraction referred to is a reflex action. That opinion was embedded in the term "tendon reflex," and it has persisted to this day—a curious instance of the manner in which a name founded upon a theory may perpetuate the theory although it has been proved to be untenable. Within a year of these observations, it was shown to be impossible that the contraction could be a true "tendon reflex." As you know, it is to be obtained in animals as well as in man. Tchirjew completely isolated the patellar tendon; he divided all its connections except those with the bone. He divided every nerve, and he found that this made no difference whatever to the effect of a tap on the tendon. That is an instance of the absolute facts of science, of the facts that cannot be got over. It alone shows that to whatever the phenomena are due they are not contractions due to the stimulation of the nerves of the tendon.

There is another interesting observation that bears on this point and can be easily demonstrated. When there is moderate increase of these contractions, if the foot is gently pressed up so as to make the Achilles tendon a little tense, the muscular contraction can be obtained as well by a tap on the edge of the tendon as on the posterior surface. If the tendon, when tapped on the side, is supported on the opposite side (by fingers against it), so that the tendon cannot move when tapped, no contraction occurs. Tap gently the tendon so supported, and then remove the support and tap again with the same force: the first tap has no effect, the second tap causes a contraction. The difference is that in the one case the tendon moves before the blow, in the other it does not. That is, unless the tap increases the tension on the muscle, it has no effect.

This fact proves that the effect must be due to the increased tension, acting either on the muscle or on the tendon itself by its nerves. But the experiment I referred to, in which all connections of the tendon were divided, shows that it cannot be due to any effect on the nerves of the tendon, therefore the contraction must be due to the increase of the tension of the muscle. The result has never been disputed, it proves that the tap on the tendon acts through the muscle and not through the tendon nerves. With this fact we are able better to perceive the significance of another. We can only obtain these contractions when the limb is in such a position that the tap on the tendon increases the tension of the muscle. If the muscle and tendon are relaxed the tap has no effect. There must be such a degree of tension as to enable the tap on the tendon to increase the tension of the muscle, or no contraction occurs.

Let me here divert your thoughts to the fact I have recently pointed out to you in connection with locomotor ataxy, the fact that the muscle nerves are stimulated by two mechanical processes—by compression and by extension.

You may remember that I mentioned to you that it is the extreme compression of the over-contracting fibres that gives rise to the pain of cramp, and that when the muscle nerves are rendered hyperæsthetic by the effect of the cramp it is by tension that the pain is produced. The pain by compression during cramp, and the pain by tension in the after-state, afford us a most instructive illustration of the two modes in which these different muscle nerves are stimulated. We are, in general, unconscious of their existence, their stimulation gives rise to no sensation. I use the term "afferent" because they do not normally influence consciousness. They are not "sensory." Nevertheless, the impulses act upon the motor centres in the cord, and, probably, also on the centre in the cerebellum influence their action by the state of tension or contraction of the muscle.

To revert to our local "phenomena;" it is thus in harmony with what we have learned in other ways that this sudden increase of tension should act upon the muscle and muscle nerves. The effect of a blow on the tendon is produced through the muscle. So far as it is reflex, it is a muscle-reflex action, and not a tendon-reflex action. There may be error of name; that is, alone, of little consequence. You may call a pen by some other name without interfering with its use, but it is different if the name involves a theory. You cannot use these phenomena, as an aid to understanding disease, if you call them "tendon reflex." At least it is necessary for every one first to unlearn the erroneous idea contained in the name, and, as a fact, a large number of persons never unlearn it and never gain the clear perception they might acquire of other phenomena of disease.

Let me now try to show you that a certain tap on the tendon is effective only when it can increase the tension of the muscle. In this patient, suffering from hemiplegia,

I press up the foot. Unless the foot is pressed up so as to make the muscles tense a contraction cannot be obtained. For it, there must be gentle tension. Hence I have called the phenomena "myotatic" or "tense-muscle" contractions. When in excess, and the muscles are made tense, the very slightest sudden mechanical influence produces the needful stimulation. I very gently tap the tendon from the side, first without support on the other side, and then supported by the finger tips so that the tap does not displace the tendon. You note that the tap, which in the first instance caused contraction, has no effect when the tendon cannot move—that is, when the tap cannot suddenly increase the tension on the muscle. You see the difference is clear. This fact, taken in conjunction with the other fact I mentioned, though the division of the nerves of the tendon prevents it, makes certain the fact that the increased tension does act upon the nerves of the muscle.

Then comes another question which is, however, comparatively unimportant. Is that contraction an immediate reflex effect of the stimulation of the afferent nerves of the muscle? It is a difficult question, and the answer depends on difficult time measurements. When the interval between the tap and the contraction is carefully measured it is found to be shorter than that which we can well conceive can be that of a reflex action. But I must ask you to allow me to pass this by. It seems to me probable that each contraction is not reflex. But this has nothing to do with the exclusion of the tendon nerves as the source of the afferent impulses. Yet I shall ask you to let me assume the theory that each contraction is not reflex, because I can better thus put before you the general principles on which I desire to insist, which help us most to understand the phenomena of disease and of health, and of which this special point is only one unessential element.

Far more important than that is the way in which the theory I am about to mention harmonises with other facts and illustrates many phenomena of disease. We only get these contractions when the muscle is in a certain state of tension, and the theory I refer to is that that tension, by gently stimulating the afferent nerves of the muscles, induces a state of the muscular fibres in which they are excessively sensitive to the mechanical influence of the suddenly increased tension of the tap. The gentle tension causes the irritability. The irritability is reflex; the contractions are local, and would not occur were it not for the irritability which is reflex.

In connection with this, two other facts are of importance. First, there is the condition called muscular "tone" or "physiological tonus," the curious state of slight contraction of the muscle by which the muscle is always adapted to its posture, and always maintains a certain amount of contraction whatever the proximity or distance of its attachments. This depends upon the spinal cord. If the motor nerves are divided it ceases instantly; it ceases instantly also if the sensory nerves are divided, and, therefore, it must be a reflex process. It is probable that this muscular tone, when a little augmented by an increased afferent impulse from tension, is the irritability which makes the muscle contract in response to the local stimulation. In most instructive harmony with this we find that whenever these phenomena become excessive, there is a tendency to tonic spasm, which is seen in such striking degree in spastic paraplegia, and, as you know, is always accompanied by such conspicuous excess of these contractions. Compare the light which that throws on the symptomatic pathology of spastic paraplegia with the absolute darkness in which the "tendon-reflex" theory would leave it; if, indeed, it were conceivable that the tendon-reflex theory could be held by any one who knows the facts.

But the name "tendon-reflex" goes on and conveys its theory to those who know nothing of the subject except the name, and thereby hinders them from all perception of the really instructive elucidation conveyed by the alternative term the "muscle-reflex action." Therefore I never use the term "tendon reflex," nor have I done so for many years. I use the term "muscle reflex," or the general term which I have mentioned, "myotatic." That word, however, has not become much used. I am, however, consoled by the different fate of another term I suggested at the same time. At the meeting of the British Association at Cambridge I was talking with Professors Clifford Albutt and Westphal about the error of the term "tendon reflex"—which Westphal (who first pointed out the importance of the symptom) had always avoided. He preferred the name "knee-phenomenon." I asked, "Why not call it the 'knee-jerk'?" Clifford Albutt so emphatically approved the term that I took the first opportunity of setting it going, and its use immediately became universal.

The contraction which follows the tap is an instantaneous one. The increased tension acts on all the fibres of the muscle; all contract. It is an instance of a simultaneous single contraction of muscular fibres. When there is maintained contraction of a muscle, it depends upon the fact that a similar contraction occurs in different fibres not simultaneously, and the contraction is maintained because there is never a moment in which a large number of fibres are not contracted. Clonic contractions are simultaneous; tonic are not.

If the muscular irritability excited by tension is considerably increased, and the tension is suddenly applied, it causes a single contraction; as this ceases, the excitability of the muscle is renewed, and, if the tension is maintained, contraction occurs. That is the reason why, by maintaining the tension upon the muscle, a clonus is produced

—a foot clonus or a rectus clonus. There is no essential difference between clonus and the knee-jerk, except that the irritability must be excessive to permit a clonus to be obtained.

The patient whom I now show you, has an extreme excess of this irritability. Before pointing out its features, I may mention the conditions under which it becomes excessive. You remember what I said recently regarding the impulses from the muscles in locomotor ataxy, and the way in which their loss deprives the spinal centres and the brain of the guiding impulses which depend upon the state of the muscles, and so disturbs co-ordination. I indicate on the black-board, in the diagram before you, a couple of muscular fibres, and the connective tissue between them, and from between them a couple of afferent nerves fibres going up, through the ganglia on the posterior root, to the spinal cord. These fibres have different destinations. One passes through the postero-external column, ascending in it to the medulla, where it ends in grey matter, whence a path continues it to the middle lobe of the cerebellum; there its impulses are combined with others to guide the motor cortex of the brain. If the nerves are degenerated, the postero-median column degenerates up to the medulla. The other fibre seems to pass into the posterior cornu, and there, probably dividing, comes into relation with the processes of the motor cells, from which a motor fibre passes down to the muscle. This afferent fibre, acting on the motor cell, is a type of the fibres which determine what I call the muscle-reflex action, that is, from the muscle to the muscle. These determine the physiological tone and determine also the excitability manifested in the knee-jerk and, when excessive, in the clonus. But the motor cells form part of the voluntary path from the brain. This descends in the lateral column, in the "pyramidal tract," the fibres of which seem to end by branches that are related to those of the motor cells; thus the pyramidal fibres act through the motor centres of the cord, which are also influenced by the impulses from the muscles. We cannot say how much these afferent muscular impulses do in determining the form and degree of the action excited from the brain. They probably have a very great influence in facilitating the arrangement of muscular contractions, especially in standing and walking. Yet these pyramidal fibres acting on the motor centre seem also to restrain its reflex activity. We should not know this if it were not for the effect of the loss of those fibres when the pyramidal tracts are destroyed. There is an increase of the muscle-reflex action and of all the phenomena by which it is manifested. As you know, in such a case we have an excessive knee-jerk, but that excessive knee-jerk does not develop at once. If the cord is divided, there is instant excess of the superficial reflex action of that from the skin, but it is some time before there is excess of the muscle-reflex action, and it afterwards gradually increases. It seems as if the function of the pyramidal fibres was to exert a continuous gentle restraint on the muscle-reflex centre. When that influence is withdrawn, there is a gradually progressive augmentation, a sort of functional hypertrophy, of the muscle-reflex action, which leads first to excessive knee-jerk, then clonus, then gradually increasing spasm, and then to the full condition of spastic paraplegia. At last there is a condition of spasm so intense as to prevent the contractions being obtained.

I shall take some other opportunity of speaking of the fallacies there are in obtaining evidence of the presence of the knee-jerk when it is normal. I wish now to show you this remarkable feature of the excess of this muscle-reflex irritability. The patient before you is a man who has had myelitis in the dorsal region of the cord. I need not trouble you with his history. His symptoms afford evidence of perfect integrity of the lumbar enlargement, except for the secondary degeneration of the pyramidal fibres, due to the effects of the inflammation higher up. They have to some extent recovered, and can perform their function of conveying the voluntary impulse, so that he is able to move his legs, but so imperfectly that he is still unable to stand without support. He has much extensor spasm in his legs, the result of the excess of this muscle-reflex action, and the degree of this excess is manifested in a feature that is not often to be observed. Not only is a clonus in the extensors of the ankle, the calf muscles, produced instantly by the gentlest passive flexion, but when he flexes the ankle himself, by voluntary contraction of the flexor muscles, the tension of their opponents sets up clonus in them. The ordinary way of obtaining clonus is suddenly to make the muscles tense, and the tension develops irritability and produces contraction without there being any interval which you can appreciate. But, as I have explained, when the first contraction is passing off, the maintained tension acts as a stimulus to another, and the time of each contraction is such that their frequency is about 6, 8 or 9 per second. When the patient flexes the joint you see there is developed a characteristic clonus, quite like that produced by passive tension. It is a feature you will seldom see, although I confess I cannot tell you why it is so rare for voluntary movement to do that which passive movement does so readily.

It is, as a rule, only in the muscles of the calf that a clonus is obtained, not only because they are conveniently made tense in this way, but also because in them this reflex action is normally very active and important. You may often get a clonus in the muscles which move the foot laterally, and also in the rectus femoris when the patella is pressed towards the foot. I now press the patella straight down towards the foot and then tap it in the same direction. You see that then a well-marked clonus is developed. When the patient is horizontal you can obtain the knee-jerk quite readily by depressing the patella and then giving a tap on the depressing finger in the direction of the patella, but not unless it is in slight excess.

Not only can a clonus be obtained in the peronei; I have even got it in the muscles of the great toe. It is often to be obtained in the flexors of the fingers in cases of hemiplegia. I have even found it in the trapezius.

With two other practical remarks I will conclude. First, remember that when a patient is in bed the best way of testing the knee-jerk, provided that you are unable to get it by pressing the tendon in the way described, is to let the hip and knee be flexed, and to let the foot rest upon your hand quite passively, and then the leg is in the most convenient position, with just the right amount of tension for the production of the contraction. This method is of extreme value in testing the knee-jerk.

The other fact is that whenever the excess is in such degree as to enable the clonus always to be obtained, that is, when the irritability is so great that the continued tension acts as a stimulus after each contraction, there is certainly nutritional change in the muscle-reflex centre—there is persistent functional excess such as implies a nutritional change. This nutritional change is important. There has been much discussion as to the significance of a definite, distinct clonus, independent of any voluntary contraction of the calf muscles, such as you can readily perceive. I am certain of this, that for one case of error due to clonus being thought to be due to organic disease, when it was really due to a so-called functional state, twenty cases of error have arisen in consequence of the clonus of organic disease being disregarded, its significance being underrated. I could give you a long list of very instructive illustrations of the errors into which disregard of the foot clonus has led distinguished physicians.

Yet one other point. In spite of the significance of such extreme increase as you observe in this patient, as proof of structural disease, it is no proof of irrecoverable disease. Never have I seen such intense excess, with such intense spasm, as was presented by a patient of whose case probably most of you have heard, and in whom there was absolute motor and sensory paraplegia up to the level of the ensiform cartilage. In spite of the absence of cutaneous sensibility there was that terrible excess of the sensibility to the deep-seated pain of muscular spasm, so that the contractions caused an agony of such intensity that the man must have died of sheer pain in three months' time. The symptoms were due to a tumour pressing on the spinal cord; the tumour was removed, and that patient has been, for years, free from the slighest trace of increase of knee jerk, or spinal symptoms of any kind. He daily walks miles, pursues an active life, and is the head of a big manufacturing firm. Ten years have now elapsed since he was in the condition I have described, and his case should impress on you the fact that no excess of myotatic irritability, and no degree of clonus or of spasm, preclude the possibility of recovery.

LECTURE XII.

A CASE OF SYRINGOMYELIA.

Gentlemen:—As the changed conditions of my work here prevent me from continuing the out-patient clinics which I have carried on for many years, I propose to try to find some substitute for them in showing you a case each Wednesday afternoon, and endeavouring to extract from its facts such information and instruction as it will yield us.

I begin to-day with an example of an uncommon disease. Let me, at the outset, warn you against the frequent error of thinking that a rare disease can only yield you knowledge which will be rarely needed. It is not so. Diseases which are uncommon cannot be studied without considering common facts and without giving increased ability to recognise diseases which are often met with.

The patient before you came into the Hospital a few weeks ago, presenting, as he presents now, two chief sets of symptoms: he has extensive loss of sensation over the hands, arms, and parts of the trunk and legs; he has also muscular weakness, and muscular wasting in the right hand and forearm, and a little weakness in the left.

The diseases in which loss of sensation is the chief symptom are not numerous. You meet with it as an isolated symptom in functional anæsthesia, in hysterical hemianæsthesia, for instance; you find it also in some cases of organic diseases of the brain which cause hemianæsthesia, and you meet with it in organic disease of the nerve trunks. But in all these forms, each kind of sensation is involved; the tactile sensibility is usually implicated even in greater degree than sensibility to pain. But in this patient—as in most other cases of the kind—the loss of sensation presents the remarkable feature that it involves sensibility to pain, and involves sensibility to temperature —to heat and to cold—but does not involve sensibility to touch. Such a combination, with the exclusion of tactile sensibility from impairment, is met with only in one other morbid process—degenerative changes in the peripheral nerves; such, for instance, as are met with in many cases of tabes, and in a few cases of alcoholic and other forms of neuritis. In these, also, you may have sensibility to pain lost, sensibility to touch persisting, but their characteristic is that they involve the extremities of the limbs, either alone or in preponderant degree, and do not extend far up the limbs toward the trunk. But, in this case, and in other cases of the kind, the loss of sensation extends over the whole upper limbs, and involves a considerable portion of the trunk. Not only does it exist on the hands and fingers, but upon the chest as far as the fourth cartilage; it involves the neck and even the back of the head as far as the cervical nerves are distributed; it involves, in front, the lowest part of the abdomen on the right side. It also extends (an unusual feature) over the front of the thigh as far as the knee; and behind—a point to which I ask your special attention—it is found over the whole sciatic area. Thus you see, by its distribution, it is entirely distinct from the only other morbid state in which this special loss is met with.

Let me show the facts I have described. I prick the patient on the palms and fingers and arms with a point, and you hear that he feels it only as a touch. So also on the chest, but you perceive that on the left side this condition does not extend quite so far down as on the right side. We examine the thermic sensibility by test-tubes containing hot and cold water. On the arms we find that he is unable to recognise either heat or cold, merely feeling the contact of the tube; he has some recognition of heat on the upper part of the chest and the back, but not of cold. If we were to examine the leg we should find that the condition is essentially the same. Note the instructive fact that the sensitiveness to cold may be lost when that to heat is not; it confirms the opinion that these two forms of thermic sensibility are subserved by different nerves.

The second symptom which he presents is the muscular atrophy in the right hand. Observe the peculiar posture of that hand; those of you who are familiar with the effects of muscular paralysis will recognise the position as that characteristic of paralysis of the interossei. The last two joints of the fingers are considerably flexed by the unopposed action of the muscles which bend those joints, the long flexors; when he attempts to straighten the hand—to extend it—you see the effect in still more marked degree; he cannot extend the middle or last joints of the fingers; he is able to extend the metacarpo-phalangeal joints by means of the long extensor, but the power of flexing these joints, and of extending the others, due to the interossei and lumbricales, is lost. This shows some weakness also in the long extensor. He is not able to extend the wrist together with the fingers. When he closes his fist, however, the wrist can be well extended, because then the special wrist extensors are active, while they are inactive in extension of the wrist and fingers together. You see, moreover, the distinct wasting of the thenar and hypothenar muscles, and the hollow interosseal spaces, in striking contrast with those of the other hand.

We will now see what the electrical irritability of the muscles is; this will tell us the state of nutrition of their nerves as well as that of the muscular tissue. You see that a moderate faradic current causes contraction in all the muscles of the left hand and forearm, but in those of the right hand no contraction can be obtained with even a strong current. There is loss of the excitability that depends on the nerves within the muscles; their state corresponds with that which exists higher up their course, and if we test the nerve-trunks that supply these muscles we find their fibres equally insensitive. They have undergone degeneration. If we apply voltaism to them they are equally irresponsive. If we apply it, however, to the muscles themselves, we obtain a contraction, because to it the muscular fibres can react; they do so, for instance, in the interossei, but only to a very strong current, and in the thenar and hypothenar muscles we get no response. There is thus a diminution of nerve and muscle irritability corresponding to the wasting, with persistence of a very slight excitability in some of the muscular tissue. It is a condition that results from, and is evidence of, a slow degeneration of the motor nerves, such as results from slow gradual damage to their nerve cells in the spinal cord.

This disease affords us a striking illustration of the great increase of diagnostic power that has come from the union of pathological and clinical observation. From its clinical characters it would not be possible to infer the exact nature of the lesion; and what may seem more strange, however familiar we might be with the pathological change, if we knew it alone, we should be unable to infer from it that such symptoms would be produced as are present in this patient, and which attend the morbid process in the majority of cases. But let us see how far a study of the symptoms can lead us. Such an extensive differentiation in the loss of the different forms of sensibility can have only one meaning. We have seen that it does not occur in degeneration of the terminations of the nerves in that distribution which we perceive in the case before us. It does not occur in disease of the nerve trunks, which causes a general loss, corresponding to their distribution. But in the spinal cord we know that the paths for pain and touch are separate, because we know that various lesions of the cord—traumatic lesions, for instance—not uncommonly involve sensibility to pain, and leave sensibility to touch unaffected; and we know also that the path of temperature must be near that of pain, because the two forms of sensibility are so frequently involved together. Therefore we can infer that there is probably a lesion of the spinal cord so situated as to implicate the paths of pain and of the thermic sense, and not that of touch. Beyond this we could not go except to infer, from the gradual progress of the symptoms—to which I shall refer later on—that it must be a lesion of chronic course. But the combination of clinical and pathological observations I have mentioned, has shown the remarkable fact that in most cases in which this combination of symptoms is met with, in any distinct and characteristic degree, the obtrusive pathological feature of the morbid state of the spinal cord is the presence of a cavity or cavities within it. The disease is that which is currently known as "syringomyelia;" it is sometimes called hydromyelia, but there seems a tendency to use the term syringomyelia to the exclusion of the other. These tendencies of nomenclature, I may remark in passing, are so strong that it is generally useless to attempt to oppose them; we have to accept that name which is generally adopted, although the term "syringomyelia," since it comes from a word meaning "trumpet," is not a very apt one.

Cavities in the cord, most of which give rise to symptoms more or less similar to those before you, are very various. The most common and the most typical is the dilatation of the central canal. It may be slight, and then may cause no symptoms; it may be great, so great as to be equal to the normal area of the cord, and by the pressure it exerts it may reduce the tissue of the cord to a narrow layer surrounding it. Lower down, there may be a normal central canal, or a central canal which is closed by a mass of nuclei, a condition which is often met apart from any disease. In other cases, with a normal or a closed canal, behind the posterior commissure there is a cavity—a slit-like cavity—between the two posterior columns, extending almost to the posterior surface of the cord. You will see examples of both forms under the microscopes. Further, we may have slit-like cavities, more or less distended and widened, in each postero-lateral column, or sometimes in one only. Similar cavities may be found within the substance of the posterior horn, extending from the front to the back of the horn. Among the sections for your examination are examples of both these changes, fissures in the posterior columns, in both posterior cornua, and in one only. Sometimes we find small slit-like cavities in the posterior part of the lateral columns, and of these also you will find an instance. We may even have a cavity, on one or both sides, at the inner part of the middle portion of the grey matter, near the base of the anterior cornu; and by its enlargement, such a cavity may come to occupy a large part of the place which should be occupied by the anterior horn and its nerve cells.

But another most important fact is illustrated by these sections and is generally to be observed. The cavities are not the only pathological feature. If you examine them closely, you will see that they are bounded by a layer of tissue different from that of the rest of the cord, a tissue resembling that which lies on each side of the central canal, and which occupies the furrow at the side of the posterior nerve roots, and which lies also in a thin layer between the pia mater and the cord, and is continuous with the neuroglia which penetrates and ramifies among the nerve fibres of the white columns. It is a relic of the embryonal tissue of the cord. It is similar to that of which the embryonic cord is composed, and which, in the process of development, slowly changes to nerve elements—nerve cells and nerve fibres. The presence of this embryonic tissue is evidence of the important fact that most of the cavities, certainly all those around which it can be recognised, and probably some others, are due to an arrest of development. They are due to the persistence of the early cavities, or to the breaking down of persistent embryonal tissue, or to both processes. This conclusion is confirmed by some other histological features which are found in the cases in which there is a distinct layer of this neuroglial tissue, and which may be sometimes found when this layer cannot be clearly seen, especially a peculiar sinuous membrane lining the cavity, or peculiar fringe-like processes extending outward from such a membrane.

The origin of most cavities in arrested development is thus evident from their features. In the posterior commissure the epithelial lining may show that a cavity is the dilated canal. The embryonal groove is, you will remember, at first closed at the posterior surface only; some cavities seem to represent an arrest of development at that stage, because the cavity due to the enlarged central canal extends backwards up to the surface of the cord. In other cases, as in one under the microscope that I have already mentioned, the closure of the original cavity behind the central canal, to form the posterior commissure, has taken place, and then an arrest has occurred, so that the cavity is between the two posterior median columns. When a slit-like cavity is in the posterior part of the lateral columns, and sometimes in the posterior horns, or in the lateral columns, the adjacent neuroglial tissue, which should not exist there, shows that the embryonal structure has failed to develop, as it should have done, into nerve elements, and, probably, most of these cavities are due to its disintegration. But a cavity in the posterior horn, or intermediate grey matter, may end at the middle line, by a rounded space in the position of the canal, and here an epithelial lining may prove that the cavity is continuous with the enlarged canal. I show you such a specimen. You may remember that, at first, before the posterior commissure is formed, there is a short extension of the canal on each side. It would seem that such cavities have arisen by the extension of one of these processes, after the union behind.

But not only does the persistent embryonal tissue remain in the amount which we should expect from arrested transformation,—it is sometimes in much greater amount, and has manifestly been increased by a process of active growth. You may perceive indications of that in one of the sections under the microscope, in which there are rounded projections into a cavity, clearly due to arrested development. The process of growth may much increase the quantity of this tissue, and even make it preponderate sometimes in an extreme degree over the cavity. A cavity may even become obliterated, perhaps quite early in life, by the exuberant hyperplasia, and a state of "central gliomatosis" may effect damage, and cause symptoms, like those that are produced by a distended cavity. The cavities formed by these processes, and likewise the overgrowth of tissue, are most common in the cervical region; sometimes, however, they extend a long way down the cord, and they may occasionally be seen in the lumbar region. The distension of the canal may also pass up even to the fourth ventricle, which may be increased in size, and the process of growth of the residual tissue may invade the floor of the fourth ventricle, and extend even as far as the upper part of the pons. Finally, as an additional evidence of the developmental and congenital nature of the changes we have considered, it is not rare to find, in conjunction with it, some developmental defect in the brain.

A few other forms of cavity are met with in the spinal cord, in rare cases, with which we need not concern ourselves just now. There may be a rarefaction of the grey substance, and a disintegration of it at some point; but such cases are quite distinct from those which form our subject to-day. They are chiefly senile or due to manifestly acquired degeneration. I believe that a similar disintegration of the white substance into a cavity only occurs in the neighbourhood of traces of undeveloped embryonal tissue. There, you may see that the nerve fibres and the interstitial tissue are alike scanty, and they may easily break down into an irregular cavity, the margins of which indicate its origin.

You can readily understand that a morbid process which presents so many variations gives rise to symptoms that differ considerably in their character in different cases. In spite of these variations, in the great majority there is the loss of painful and thermic sensations. We do not yet know what mechanism in the morbid process gives rise to this peculiar symptom; how far it is due to simple compression of the conducting tracts that convey these forms of sensibility, or how far it is due to damage to the fibres of the posterior commissure, by which we must assume that these sensory impressions cross. Again, when there is a cavity in the posterior horn, we can hardly conceive that the nerve cells, to which probably the nerve fibres first convey these sensory impulses, are not also damaged. These facts render it mysterious how sensibility to touch can so frequently escape. There are cases, however, in which all forms of sensibility have been affected. The affection of sensibility is commonly met with in the arms and parts of the trunk, very seldom, as in this patient, in the leg, though the legs are not seldom weak and present the characteristics of spastic paraplegia, a fact which we can understand when we consider the pressure which the pyramidal tracts must undergo when there is great distension, or overgrowth of the embryonal tissue, or cavities in the lateral columns themselves. Muscular wasting, such as we see in this patient, is common, in one or both hands, sometimes chiefly in the shoulder muscles. It is apparently due to the damage to the anterior cornua by the enlarging cavities or compressing growth, and on the locality of these its position depends. Other trophic changes are sometimes met with, the most frequent of which are lividity at the extremities of the hands, and interference with the growth of the nails; occasionally, especially where there is muscular wasting in the upper part of the arms, there may be changes in the joints, especially in the shoulder and elbow, resembling those of tabes, and differing from those in chronic rheumatoid arthritis by the fact that the thickening takes place chiefly outside the capsule of the joint and not within it. Other symptoms, which, however, are rather rare, are the occurrence of ulcers and other changes in the nutrition of the skin of the fingers. More frequent are those which are due to the ascension of the affection to the medulla. The sympathetic may be involved by the lesion, so far as concerns the eyes, when the mischief extends to the upper cervical region of the cord; and hence changes of the pupil-contraction have been sometimes noticed. The bulbar nerves occasionally suffer, and also those for the external ocular muscles, the nuclei of which may be reached by the distending force. In this patient there is weakness of the external recti, probably due to the distension of the fourth ventricle. There is also nystagmus, which has the same significance, and is not at all rare in this disease.

One other pathological fact should be mentioned to complete our general survey of the more salient features of this strange malady. In several cases degeneration has been found in the peripheral cutaneous nerves. It has been chiefly looked for when there have been trophic changes, and has then been constant. It seems scarcely likely that the special sensory loss is due to this, since the degenerated nerves are of the extremities, and the loss of sensation is so extensive.

Such a malady, so peculiar in its manifestations, is not likely to be mistaken, provided only that its features are known. There are indeed only two diseases with which this is likely to be confounded, provided you know its features. One of these is chronic cervical pachymeningitis, in which there is much thickening of the cervical dura mater, and damage to the nerve roots. In such a case you have wasting and loss of sensibility in the arms, but the loss of sensibility involves all forms, is less extensive, and is preceded by much pain. These are distinctions which seldom fail. The other condition is one which cannot be entirely separated from the disease we are considering, because it seems to be, in part at least, due to the same morbid process; it is the affection which has become known under the name of "Morvan's Disease," or "Painless Whitlows," in which whitlows form on the fingers, and produce ulcers which do not heal for a long time, sometimes indeed only when the finger-ends have been lost. The painlessness is due to the loss of sensibility to pain, commonly in conjunction with loss of sensibility to temperature, but without loss to touch. I may mention, as a striking instance of the analgesia accompanying this affection, that the shoulder joint has been resected without pain, although no anæsthetic was used. "Morvan's Disease" is found to be due to cavities in the cord, similar to those in the condition of syringomyelia, but combined, apparently in all typical cases, with extensive chronic degeneration, or degenerative neuritis, of the peripheral nerves. Its chief distinction is the presence of these painless whitlows, and there is usually a greater limitation of loss of sensibility to the lower portion of the limbs. This is thus rather a variety of syringomyelia than a separate disease.

These, then, gentlemen, are the main features of syringomyelia; and the chief characteristics by which it is manifested. It is a disease for which, you will be prepared to hear, we are able to effect little by medicinal treatment, though we may be able to give temporary relief to its symptoms. This patient complained greatly, on his admission, of distressing pain in the head, probably due to increased pressure within the cavities of the brain. This symptom has been greatly relieved by antipyrin. Electrical treatment may maintain for a time the nutrition of the muscles, but when their wasting depends on a morbid process that destroys the motor cells, we cannot hope for any regeneration of the nerves to reward our efforts; we cannot anticipate any restoration of the path by which the will can act upon the muscles. But the future may perhaps have in store a ray of light to lessen the present practical gloom. Where the physician fears to tread, the surgeon sometimes steps in with the best results, and we must humbly bow in grateful recognition of the good he has succeeded in effecting in recent years, the lives he has saved, and the hindrance to the progress of disease which he has produced, the relief he has given, when we could not save or relieve or retard. But I am not aware that the surgeon's skill has yet been applied to this affection.* It seems to me that it is quite possible that, in some cases, surgery may be able at least to effect an improvement in

---

* Since this lecture was delivered I have found a report of one case that was operated on in the manner suggested, without ill effects, with transient amelioration, but no permanent benefit. The case was, however, unfavourable, because the chief symptoms were in the legs, and cystitis ultimately caused death. (Abbe and Coley, *Journ. Nervous and Mental Dis.*, July, 1882.)

the condition of the patient, and a retardation of the progress of the affection. The exposure of the cord would readily enable him to ascertain whether the process was the distension of a cavity by liquid, or the presence of a morbid growth. A central growth, extending high and low, could not be touched, but it may prove to be different when the chief element in the lesion is a distended cavity. The draining away of the fluid contents might enable a cavity to contract, and when the amount of circumferential growth is not too large, the relief from the slow compression of the adjacent structures, would probably permit their recovery to a considerable extent, because no effects pass away so readily as those which are produced by slow compression, when that compression has been removed.

The malady we have been considering is thus one of those developmental diseases on which we, as practitioners, can only look with feelings which are the reverse of gratifying. It must be so. Such affections are the result of morbid tendencies, perhaps inherent in the germ, certainly beginning in the embryo, continuing in childhood, and becoming active and aggressive in a later period of life. Their activity seems often to coincide with the completion of the process of growth; but over the morbid tendency we are as powerless as we are over growth itself, or as the monarch of old was over the waves of the advancing tide. The advance of these diseases is sometimes rapid, sometimes slow; sometimes, like the waves on the shore, they may seem for a time to give no evidence of onward progress; they may even seem to recede, but it is only for a time. Before long a fresh advance carries their effects beyond the highest point they had before attained. But here, alas, we find that our simile fails us. The tendency of these diseases knows no retrocession. It is a tide which flows, but never ebbs.

LECTURE XIII.

THE TREATMENT OF MUSCULAR CONTRACTION.

Gentlemen:—You are familiar with the fact that consequences of disease are sometimes more serious than the maladies from which they result. I do not now refer to the cases in which one definite disease is produced by another, such as the paralysis due to the poison that is the result of diphtheria, which causes such grave after-effects. I desire to draw your attention only to some simple results of disease, results that are the necessary consequences of the primary affection,—which we may almost describe as the natural effects of unnatural states. But, although they are normal consequences of abnormal conditions, they reach, by their simple increase, a degree which constitutes so definite a disorder that we are inclined to regard it as a morbid process. A morbid process it is, so far as its effect is concerned, although it is not a morbid process in its nature. It is disease only because it is an excess—because, in the absence of the normal limitation, it passes on to a condition which disorders normal action.

This being the nature of the contractions of which I am about to speak, I must point out to you their causes and mechanism, that you may understand them, and that you may discern their nature when you encounter them; that you may know what they mean, and that you may know how to remove them—to remove them if it may be, to lessen them if they cannot vanish, and, above all, to anticipate their occurrence, to discern the conditions in which they will develop and yet in which they need not develop, to prevent them in the strictest sense of the word—to come before them, to stand in their way, to *obviate* them.

Such a secondary process, arising from the unrestrained action of processes which are normal, is the cause of many of the results of disease which are most difficult to remove. Indeed, removal is only possible when the normal restraint can be renewed. The process of diminution even then is very slow. The renewal of restraint is gradual, and has to act on a tendency that has become far greater than normal, and, moreover, has a tendency to increase. This has first to be arrested before diminution of the abnormal degree can be achieved, and the normal restraint re-established, if its recovery is adequate.

I use these general terms because the conditions described, and the influences referred to of restraint on the one hand, and excess on the other, may be perceived in many morbid states, and their discernment enables us to understand better many processes and many consequences which outlive their causes. They often seem like independent maladies when they are only properly to be regarded as runaway vitality. For they depend really on the normal power of living structures to adapt themselves, by their nutrition, to conditions different in character or only in degree from those of health.

This process, as I say, may be met with in many affections, but in none more clearly than in the condition to which I want to direct your attention to-day, the muscular contraction met with in many paralysing maladies of the spinal cord and of the nerves.

The tonic contraction of muscles that occurs in organic disease may be divided into two chief forms, easily distinguishable, and important in their significance—that which yields to an attempt to extend the muscle, and that which cannot be overcome.

I propose to consider to-day only the form that cannot be overcome. The calf muscles are often thus contracted as I show you in this patient, who is recovering from multiple neuritis. When the foot is pressed up by the hand, placed against the sole, the calf muscles resist, the contraction yields a little, and the foot is brought almost or quite to a right angle with the leg; but however long you continue the pressure the foot goes no further. That is the contraction which depends upon tissue-changes in the muscle, causing it to be structurally shorter than normal. It is of very great practical importance. You may remember observing it in the boy with idiopathic muscular atrophy whom we examined last week. The first point for us to consider is to what it is due.

It is met with especially in muscles that are less paralyzed than their opponents, and is most common in the calf muscles, which extend the ankle joint, when the flexors in the front of the leg are paralysed. But we also meet with it when both sets of muscles are paralysed, and also in cases in which there is no paralysis, provided extension of the foot on the leg has been maintained for a long time. It occurs also in the flexors of the hip and of the knee if, from any cause, the patient lies in bed for a long time with these joints flexed. If you scrutinise the conditions in which it is met with, you will find that it is due especially to posture—that even paralysis is only effectual by inducing a certain posture. You know that there is an extraordinary capacity of the muscles to adapt themselves to posture. Whatever the position of a limb or joint, the muscles must always continue in a certain state of slight gentle tonic contraction—"physiological tonus" or "tone" as it is called. If a limb is moved passively, the muscles which are relaxed do not wrinkle up; they remain contracted in exact proportion to the approximation of their attachments, and the muscles which are extended elongate, but remain in similar tonic contraction in exact proportion to the increased distance between their attachments. That capacity for adaptation to posture is evidently one phase of the condition which physiologists call "tone," and which depends upon the connection of the muscles with the spinal cord. From the nerve cells, in the anterior cornu, motor fibres proceed to the muscle. From the tissue between the muscular fibres, other nerve fibres pass up to the posterior cornu of the cord, and are connected, by a mysterious interlacement of fibrillæ, with the branching processes of the cells of the anterior cornua. On the integrity of that arrangement this adaptation to posture and this muscular tone seem to depend. In consequence of that, if the attachments of a muscle are permanently and constantly approximated in consequence of posture, the changes which, as you know, are always going on in every structure and every tissue of the body, alter the muscular fibres and the interstitial tissue, in accordance with the diminished length, so that after a time you are unable to extend the muscle— unable, that is, by any attempt which you may make at the time. We meet with this in cases of unequal paralysis, and also in cases in which one posture is constantly adopted. This effect of paralysis may be always traced to its influence in causing a certain position of the parts. That is the great cause of the various forms of talipes that result from the unequal palsy of muscles in infantile paralysis. The most common of these forms, that to which I want specially to draw your attention as typical of the others, is the contraction of the calf muscles when those in front of the leg are paralysed.

We meet with it also as the result of posture, determined by pain, and also of posture determined by simple comfort. It is probable that when there is infantile palsy with this sequel, the contracture of the muscles is facilitated by the fact that they also have suffered a little, that there are slight interstitial changes, nuclear multiplication, with increased development of new fibrous tissue which contracts. But remember it is not a consequence of the palsy; it occurs without it—I mean it is not a consequence of the slight palsy of the contractor muscles; it is the consequence of the persistence, for a long period, of the state that results from the capacity of the muscle for adaptation to posture.

It is a grave inconvenience; it hinders the recovery of use of the legs for standing and walking in patients who have otherwise gained sufficient power. It often needs division of tendons in order to permit the wrong posture to be rectified; but it does not always need it. In many cases, unless there is shortening of the leg, the persistent, long-continued extension involved in the act of walking elongates the contracted muscle to its proper length. And that is the case in all conditions in which perfect recovery of the damaged structures causing paralysis is possible. It is conspicuous in adults—in whom no hindrance to growth can occur. In multiple neuritis, for instance, we have much greater palsy in the muscles in the front of the leg than in the sural muscles. The latter contract, in consequence of the posture permitted by the weakness of the flexors of the ankle. It is a disease which passes away under good conditions; and however great that contraction, it always in time yields to the attempt to stand and walk. Tenotomy is never needed. Tenotomy is only indispensable when the condition is due to a state from which recovery is impossible. In infantile paralysis it is frequently needed, because infantile spinal paralysis depends upon destruction of the grey matter of the cord; recovery is impossible then, and without return of power in the opponents the contraction that results cannot be removed.

But that which is difficult to remove may be prevented. The great cause of this contraction of the calf-muscles,

A METHOD FOR PREVENTING CONTRACTION. 193

permitted by the affection of the muscles in the front of the leg, is the action of gravitation in determining the posture of the foot as the patient lies. The foot "falls," *i. e.*, always sinks into a position of extension. The calf-muscles become shortened, first by active, extensible contraction, but this becomes fixed by nutritional changes, and ultimately is absolute by structural change. But the falling of the foot need not occur. It is a point of great practical importance, and is almost entirely neglected. There never need be this shortening of the calf-muscles from loss of power in the anterior tibial muscles. It is only necessary to keep the foot always up, and the contraction cannot occur. It may be kept up either by some support below the sole, such as a plank or a thick sandbag, or else by some traction of the foot. Support from below cannot, however, be maintained in adequate continuity. I have tried every form of support that could be devised, and have been compelled to fall back on traction. The force exerted need not be great; a very gentle elastic traction, long continued, will keep the muscles adequately extended, and prevent the occurrence of this most troublesome result. But the contrivance of such traction is a much less simple matter than it may appear. There must be a place from which it is exerted. It must come from some part near the knee. However gentle it is, by long continuance it becomes annoying, then painful, and ultimately unbearable. It can be borne, indeed, for a very little time when there is tenderness, as there is always in the condition in which it is most needed—cases of multiple neuritis. I tried traction from the upper part of the bedstead, but the changing posture of the patient made this useless. At last, after many trials of various forms, an idea occurred to me of a plan by which the traction could be exerted from the upper part of the leg near the knee, which has proved perfectly successful. I described what I desired to our excellent resident medical officer, Dr. Whiting, who speedily carried it out in perfect detail, with a success which is most gratifying. I show you here the apparatus.

There is a leather sheath for the leg, almost meeting, laced together in front by a cord passing round the hooks. It may go above the knee or stop short of it. There is a similar sheath for the foot, and the two are connected by an isthmus of leather about an inch wide. From rings at the

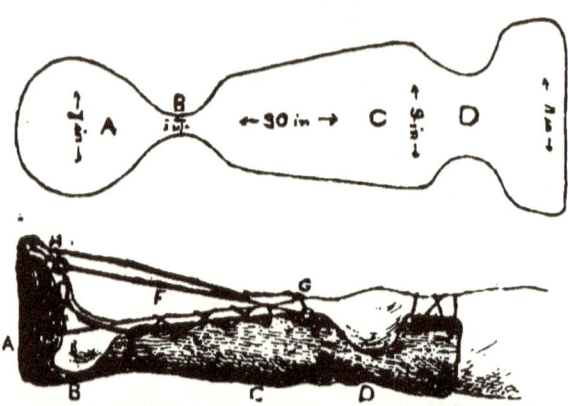

The upper figure represents the shape of the brown paper pattern, and of the leather, cut in correspondence with it. The dimensions are approximate, varying with each leg. D is the narrowing for the knee, that the leather may not press on the bone at the sides. B is the excavation on each side for the ankle with the narrow connecting piece of leather, which is the special feature of the contrivance. This, 1 to 1½ inches wide, and no longer than is needful, affords, lengthwise, adequate support to make the pull from above act down on the foot-piece on which the pull acts. At H and G are the rings for the cords. The cords are here shewn only as going to the lower rings; if there is contraction of the knee they should be carried through these rings and fastened to the other rings above the knee. If there is no contraction of the knee it is not necessary for the leather to extend above the knee.

end of each sheath cords extend, in which India rubber is interposed. A strong, common India rubber band answers perfectly well, for the elastic force, as I said, does not need to be great; it needs only to be constant. The slightest constant traction is effective, but constancy, under varying mechanical conditions, can only be secured by elastic traction. It is remarkable how well the muscles bear such traction if only it is gentle. To the more sensitive skin it soon becomes unendurable, but the impressions from the afferent muscle nerves only act on consciousness, so as to cause "sensation," when excessive, and even long-continued gentle extension seldom causes muscular discomfort.

The new feature which it embodies is this. The part of the apparatus for the leg *from* which the traction is made is continuous with that for the foot *on* which the traction is made. The two are continuous by a narrow piece of the leather, which will bend so as to permit the traction to increase the flexion of the ankle, and so to overcome the shortening which has occurred, and which yet presents longitudinal resistance. This resistance is enough to make the source of traction, and that on which it is exerted, effectively one. Instead of the pull being from the leg itself, it is from the apparatus that the force acts on. Continuous traction from the leg itself—which, however slight, becomes unendurable by persistence—is thus avoided. The traction is exerted from a part of the same structure as that on which it is exerted. This piece of leather, narrow and short as it is, affords sufficient support, and no pressure is caused on the leg itself.

Another advantage of this apparatus is that it is easily made by any practitioner. A piece of leather must be cut out, of the shape which I now show you—two ovals connected by a narrow piece. It is best first to take a sheet of brown paper and cut out a pattern to fit the leg and foot of the patient. The leather must then be softened in water, warm at the last, so that the leather is warm when it is applied to the leg. It must be bound on the leg by cord or bandage, and then allowed slowly to dry and harden. The leg should be swathed to prevent chill from the drying. If thought desirable, oil silk may be placed next the skin. When dry and hard it is taken off. It has only to be lined with wash leather, and any shoemaker will insert in it the necessary hooks and rings. A piece of cord and an elastic band complete it. It may be made in rough but perfect efficiency, with holes and tapes or cords, and a penny elastic band for traction.

Mr. Hawksley, the well-known surgical instrument maker of Oxford Street, has made a copy of this apparatus, finished off with all due skill, and therefore more sightly than that which I show you, although it could not be more effective. He has been good enough to place it in his shop for any one to examine. As the only apparatus known to me that is really effective for preventing a grave evil, and even removing it when it has occurred, at least when it has not reached an extreme degree, and as one that is easy to make in perfectly serviceable form, I think it deserves to be widely known. It should be adopted in all cases in which a patient has to be kept in bed, and the weight of the foot leads to persistent extension of the ankle-joint, whatever be the cause of the condition, if there is the slightest evidence of commencing contraction of the extensors, or if there is even reason to suspect that it is likely to come on.

But I have found that it produces another effect which I did not anticipate. When the leather is carried a little above the knee, as it is in the form shown in the figure, any tendency to flexion of the knee is counteracted, and with it flexion of the hip. When a disease that can be recovered from leads to such flexion, prevention is of great importance. When it occurs as the sequel to extensor spasm, treatment has little chance of doing good. The flexor spasm then says unequivocally, "Hope is at an end."

LECTURE XIV

THE INFANTILE CAUSES OF EPILEPSY.

I.

Gentlemen:—Many pleasant and useful memories remain to me of the two years' apprenticeship to a country surgeon by which I commenced the study of medicine. One incident often comes to my mind in connection with the subject of infantile convulsions. I remember spending some hours in a cottage, one summer afternoon, watching a child in "teething fits." The child was kept continuously in a warm bath. I do not know that my presence there was of much service to the patient, but it is of service to me now. It brings before me, very forcibly, the changes which have come to our conceptions in the thirty years that have elapsed. Fits, associated with dentition, were then always due to "cerebral congestion," excited by the irritation of the backward teeth, and it was necessary to keep the blood in the vessels of the skin that there might be less in the intracranial vessels. I remember well how the colour of the skin testified to the continuous action of the process, and yet how the convulsions continued without the slightest alteration, although the "derivative" influence of the bath was so marked. It was, indeed, some years before this time that the true relation of convulsions to dentition was made clear by Sir William Jenner. But knowledge filtrates slowly through the profession, or at least its diffusion was then slow in rural districts, even among intelligent practitioners. The lectures on "Rickets" of Sir William Jenner not only marked, but made a turning-point in professional knowledge. The facts he demonstrated have now for long become general knowledge, and their chief source is lost to sight. The perusal of his lectures cannot, however, be too strongly recommended, and, happily, their forthcoming reprint will make it easy.

No one now dreams of associating the common convulsions of retarded dentition with congestion of the brain, although a direct influence on the brain by the irritation of the process of teething is still sometimes thought to be the sole element in causation. In general, however, the influence of the process of the eruption of the teeth is relegated to its proper place, as merely a possible excitant in a few cases.

I hope it will not be very long before the progress of medical knowledge, and the improvement in medical education, makes "congestion" everywhere a morbid state that needs definite discernment, and not merely the equivalent for a negation. This may be, perhaps, too much to hope for. The assumption of "congestion" of this or that organ, of the liver in dyspepsia, or of the brain in hypochondriasis, is so easy; its assertion is so precise, so satisfactory, and it enables treatment to produce such a definite result, that the assumption will not readily be regarded as a thing to be proved. As far as can be judged, its end is not yet. Of course, I am speaking only of the active congestion, which is constantly invoked to explain that which is not understood, not of the passive mechanical congestion, which is quite a different process, and is ignored when it exists as frequently as active congestion is assumed when it is absent.

The service which Sir William Jenner did was to demonstrate that the condition of rickets is a general retardation of development, with various secondary, necessary results, and that the association of convulsions is with this general state, of which the backward teething is only one of the manifestations.

The convulsions are a consequence of the retarded development which occurs so often towards the end of the first year. This epoch is one in which development is readily checked. Growth, indeed, goes on, but the progressive development of the tissues is hindered; their elements are still produced, but do not present the progressive development that should coincide with general growth. It is the epoch at which the character of the food-supply undergoes change, often unwise in suddenness, or does not undergo the change that is natural. It is a period also, when much functional capacity passes into functional use. Any general illness or prolonged digestive disturbance may have, as a result, distinct signs of this malady, but in considerable degree it is the result of preventible causes, which it would be foreign to my present purpose to discuss.

The influence of rickets in causing such fits is of great importance in connection with epilepsy, because there is no doubt that such convulsions are a frequent element in the causation of that disorder. They leave behind a residual disposition to the like morbid action, which may be continuous in its results or may become active at a later period of life. Between the cases in which dentition convulsions have ceased after a few months and convulsions have recurred as epileptic fits at puberty, on the one hand, and, on the other, cases in which the infantile convulsions did not cease, but continued into life-long epilepsy, we have every gradation. There are cases in which the attacks went on for a few years, ceased for a year or two, and then returned, and cases in which the interval of freedom was five or six years. Indeed, every variation of interval is met with. It is thus impossible to doubt that the dentition convulsions are a definite element in the causation of epilepsy.

So constant, moreover, is their association with the defective development which we call rickets, that it is impossible to doubt that the prevention of rickets would have a considerable influence in the prevention of epilepsy.

One question to which I desire especially to direct your attention is this. How does the state of defective development cause the liability to convulsions? I believe that they can be adequately explained by the simple fact of retarded development. This is conspicuous in all parts of the system. The morbid conditions that are characteristic of rickets can be, for the most part, explained by the perversion of nutrition, which is a necessary result of the retardation of the process. In the nervous system we know that, at birth, the state of the various structures differs, that some are structurally perfect, while in others a considerable amount of developmental change has still to occur before full functional capacity can exist. It is not necessary to describe the facts in detail, because the general character of the relation is alone important. The nerve structures which are lower in function, and in part in position, are developed before the higher. This is true of the motor elements and reflex centres. The lower centres are under the control of the higher, and their activity in early infancy, when the higher centres are less developed, is manifested in the restless, aimless movements of the child, and the conspicuous activity of the reflex processes. At birth these are developed to such a degree as to enable the mouth to suck, and the thorax and larynx to co-operate in producing too-familiar sounds. The process of development goes on, and as structure is complete, function becomes established in more perfect degree. But the establishment of perfect function follows structural completion, and it is long before the influence of the relative development ceases to have an influence on the action of the nervous system. Hence it is that all through infancy and early childhood reflex processes are so active. Their early predominance is only slowly reduced into subordination as the function of the higher centres becomes established in effective degree. It must be remembered that the motor centre in the cortex of the brain, from which impulses proceed directly to the spinal cord, is a centre relatively low, and its control by higher structure is a comparatively late event. Not only does this relativity of development and function explain the excessive spontaneous and reflex movement of infancy, but it explains to some extent the readiness with which the motor cortex acts in the sudden, violent manner which gives rise to convulsions. It is sudden and violent, although incomparably slighter than the convulsions of later life. It is important to remember the fact. Without doubt many of the attacks in infancy which consist only of tonic spasm, would in an adult consist of violent clonic spasm as well as of tonic spasm. The same activity in the well-developed centres for the larynx gives rise to the familiar " laryngismus stridulus."

The hindrance to development, which acts upon all structures, of necessity has most effect upon those which are least complete. Where structure has become perfect, and function is established and is being fixed by use, its influence has little effect. Where structure is not yet perfect, and where it has just become perfect, but function is only just assuming its adequate form, the retardation has an influence in proportion to the imperfection of structural and functional development. It must be remembered that the retardation has an influence on growth and nutrition, and the effect on the finer nutrition which is related to function may be considerable, even after structural growth is at an end. Hence the relation of the lower centres to the higher, the excessive activity of the lower, which is so conspicuous in early infancy, is reproduced by the influence of rickets. The excessive activity of early life, moreover, becomes definitely morbid insubordination. The greater capacity for functional action of the lower centres needs the proportioned control which should come from the due development of the higher, and this is wanting. Hence the uncontrolled activity is manifested by morbid action, possibly contributed to by the general failure of nutrition, which involves a less perfect form of action even in the centres in which development is complete and function established. Thus we can understand, without difficulty, the morbid action of the reflex centres which causes such a state as the contracture in the hands and feet and the laryngismus stridulus of rickets, and also the occurrence of the characteristic convulsions which probably depend upon the lower cortical centres of the brain, those of the motor region.

With the restoration of the normal process of development which comes from the treatment of the general state, the higher centres acquire increased control, and, in most cases, the tendency to convulsion ceases. But the sudden spontaneous overaction has left its residual effects. These may be so strong that there is no cessation of the fits, or only a brief cessation, or a recurrence under favourable conditions. Thus we have the series of cases to which I have already referred, which demonstrate the relation of such convulsions to ordinary epilepsy. The residual influence, with its consequent tendency to persist or to recur, is certainly influenced by the mysterious functional state which is due to inheritance. The precise nature of this we do not know; we can only guess it from its effects. It seems to be a defect in the chemical composition or the molecular structure of the nerve centres, in consequence of which nerve force is liable to be released without the stimulus which alone should excite this liberation. The tendency is seen in its special and isolated operation only after infancy is passed, but during this period it seems to co-operate with other conditions, and seems to increase the relative liability to convulsions in rickets. A slight degree of hindrance to development suffices then to give rise to the convulsions. Still more markedly does the inherited tendency dispose to the residual disposition to recurrence which is left behind by every fit, to which the persistence of convulsions is due, and still more again does it dispose to a recurrence, in consequence of this residual effect (which never entirely ceases) at some period in later childhood or in youth.

All parts share in the defective nutrition and in the hindrance to their development. The bilateral symmetry of the frame involves the equal affection of both sides, and hence it is that the convulsions which are due to these causes are general. This is an important fact, as I shall hope to show in the next lecture. The persistent convulsions which may carry on the malady from infancy to adult life are of the same general character. Often, indeed, the establishment of the control of the higher centres, when it does not arrest the process of discharge, modifies its form in a way we cannot yet understand, and attacks of "minor" character replace the convulsions, to be succeeded, at the time when epilepsy is most prone to develop, by convulsive attacks of greater severity. The convulsions of rickets, as I have said, in the slightness and tonic character of the spasm in a large number of cases, are nearer to minor than to major epilepsy. This fact is important because attacks of *petit mal* may succeed the convulsions of rickets and be unnoticed until spasm is again added.

I need not say anything of the treatment of these convulsions, because it is well known to you all. But one point it may not be superfluous to mention. The bromide which is given to check the tendency to convulsions while the general treatment is re-establishing the proper course of development, should not be too hastily discontinued when the fits cease. The residual tendency must be remembered, and this may be lessened, I believe, by continuing the treatment for some months after the attacks have ceased, and the time should be still longer when there is reason to think, from the fact of inheritance, that this tendency is likely to be profound.

LECTURE XV.
THE INFANTILE CAUSES OF EPILEPSY.
II.

Gentlemen:—The convulsions which we considered in the last lecture as a cause of epilepsy—those that occur during the period of dentition and are the result of the hindered development of rickets, are general. The later convulsions of epilepsy, which are developed from them, or disposed to by them, are also general. They are the characteristic convulsions of what is called "idiopathic" epilepsy. The convulsions of rickets are moderate in severity, and when they are continued into epilepsy, the attacks often become slighter, until they are merely minor attacks with loss of consciousness; there may be only the slightest indication of spasm, or there may be no trace of spasm after a time. Such minor attacks may continue for a few years, and then again muscular contractions may be combined with the loss of consciousness; increasing in severity, the seizures gradually assume the features of severe epileptic attacks. But the practical lesson to be learnt from the facts which we have considered is this: in a case of epilepsy which dates from convulsions in the period of the first dentition, moderate in degree and general, whether the attacks have been continuous or interrupted by a few years' freedom, you are justified in considering that the case is one of idiopathic epilepsy, founded on the convulsions of rickets.

But there are two other forms of epilepsy in which the

disease can be traced to infantile convulsions. The first of these is especially important. It differs from that we have considered in three features. The epileptic convulsion is one-sided, at any rate when moderate in degree, and if there is an aura, it is in one limb or in the face on one side. The spasm can generally be observed to commence locally in the hand or face. When both sides are convulsed, one is affected before the other, except in the most violent attacks. In these there may be no perceptible interval between the affection of the two sides. Inquiry regarding the character of the infantile convulsions will usually elicit the fact, when their character is known, that these also were confined to one side. The second feature is that the first attack of infantile convulsions was of great severity. Often, indeed, there was a series of convulsions, one after another, for several hours. Sometimes such a series was the only infantile attack; often initial attacks were followed by others for a month or two, which ceased to recur when a few years had passed. Complete continuity of occurrence from infancy to adult life is less common in these than in the cases first considered. The third feature is that the first convulsions often occur during some acute illness, or soon after a fall, or in a state of general physical prostration. This is not always the case, nor is it always true that there are several consecutive convulsions at the onset. But the first fit is seldom slight in degree, and, if the facts can be accurately ascertained, the first fits are known to have been unilateral.

The unilateral character of the convulsion means unilateral instability of the motor structures of the brain. The commencement in one limb means local instability in a certain part of these motor structures. This part differs considerably from the rest in its morbid functional tendency. Such local change excludes a general cause, which would act on the whole brain. The suddenness of the onset indicates a sudden development of the instability—that is, a sudden change at one spot. This is, however, equivalent to the assertion that the convulsions are due to a local lesion in the motor structures, because a sudden change means organic disease. Without doubt, this is the fact in the class of cases we are considering. There is organic disease at the cortex of the brain. Its occurrence is indicated by the initial convulsions. The effects of the process on the adjacent tissue, which is slightly changed, is such as to induce a permanent alteration in nutrition and function. In consequence of this, the structures "discharge" from time to time. The disease may be either in or only near the motor region. If it is in the motor region, and is more than minute, there is loss of power—hemiplegia—the amount of which depends on the extent of the lesion. It may be such as to be persistent through life, or it may be transient and pass away in a few months. When the disease is only near the motor region, any initial loss of power may be too slight to be recognised. In an infant slight weakness on one side usually escapes recognition, and considerable paralysis may be unnoticed if the child is at the time prostrate from some general illness. When there is no lasting palsy, and when there is no history of initial palsy, we may nevertheless obtain some indication of it. You are familiar with the aspect of cases of infantile hemiplegia, you know that they frequently present a peculiar inco-ordination of voluntary movement, especially conspicuous in the peculiar ataxy of the hand. When this is present in the arm, you will seldom fail to discern that in a slight smile there is more movement of the face on that side than on the other. You may often detect this when there is no other indication of hemiplegia. Corresponding to the side on which the convulsions begin, or to which they are limited, it is of very definite significance. Its significance, indeed, is far greater and more trustworthy than you would conceive from the trifling nature of the symptom. It is only in a minority of these cases that we find persistent hemiplegia. The tendency to "discharge" in the motor structures is due to damage; the hemiplegia is due to destruction. The tissues which are destroyed cannot give rise to convulsion. Hence, when the damage is too slight to cause persistent loss of power, or is situated outside the motor region but near it, so that the nerve elements in the motor region are damaged but not destroyed, the tendency to convulsions is met with in chief degree.

The actual lesion in these cases is still a matter of conjecture to an almost strange degree. Perhaps, however, it is not so strange. The lesion is comparatively slight, and vitality in early life is strong and has a marvellous power of resisting the influence of local disease, readily as it often yields to a general morbid state. Although the motto of the Pathological Society asserts that death is not silent, the truth is relative. Death has her secrets still, and, when death does not unfold the closed roll, the secrecy often baffles the most careful scrutiny. Not only is scrutiny baffled, but effort is tantalised, and the region of inference is a field of contest. Hence, opinions differ widely as to the character of such sudden lesions. Doubtless we shall not have to wait much longer before facts assist us. At present the opinion which seems to myself to deserve most weight is that there is a sudden occlusion of a small surface vein by clot, with the consequent intense congestion and hemorrhagic softening of the region of the cortex. In the softened region the nerve elements are destroyed; an indurated contracted area ultimately represents the disease. There is not usually the actual softening that is met with in arterial occlusion, or in traumatic injury, but on the margin of the chief destruction, in every form, there is a region of slighter damage, and it is, no doubt, from this that the discharges proceed.

THE RESIDUAL DISPOSITION.

Remember that the function of the motor nerve structures depends on their ability to release a considerable amount of nerve force, without any appreciable interval of time after the stimulus reaches them. There must be a large amount of "latent energy" in the structures, which the stimulus at once releases as active force. In the structures in which the nutrition is deranged by the previous process, the energy is not retained in its latent form until the proper stimulus releases it; it accumulates and escapes without any influence, that we can recognise, acting upon the structures to cause the release of the force. This we call "discharge." The process, once set up, spreads widely in proportion to its energy. If this is slight, it is limited to the part in which it began; if greater, it spreads through that side of the brain; if greater still, to the opposite hemisphere. And note this—wherever it occurs, wherever it spreads—in every place the discharge leaves behind it a residual state disposing to its repetition. This residual disposition is, in health, the foundation of all our habits, of all our education, of all our acquirements, muscular and mental. It is the physical basis of memory, and it is the basis of such functional disease as that which we are now considering. It facilitates the spread of the discharge, so that the convulsions after a time, become general more readily than at first. Remember this, because we shall presently see that it has an important practical application.

Lastly, there is yet a third class of cases in which epilepsy has its origin in infancy. There are cases in which the first symptoms can be traced back, not only to infancy but through infancy—back to the earliest period of separate existence. They are cases in which there were convulsions during the first two or three days of life, or, at least, convulsive twitchings and other indications, such as difficulty of swallowing, of grave impairment of the brain. These cases occur in the children who are the "first-born," who enter the world with greater difficulty than those which follow them, and whose birth, as you can generally ascertain, was long and tedious, often needing the aid of instruments to terminate it soon enough to save the new life. You know that such children pass out of infancy into an imperfect physical childhood, with weakness of the legs, sometimes also of the arms, with irregularity of the movements of the hands, and often also with convulsions. You have heard the cases described, and they have doubtless been pointed out to you as examples of the condition for which I have proposed the name of "birth-palsy." The symptoms are the result of damage to the cortex of the brain, commonly the effect of meningeal hemorrhage. When the damage is slight there may be no symptoms of paralysis, but its effect may be manifest by the occurrence of convulsions, either in childhood or in later life. Do not forget this because, although it is not common for epilepsy to be the sole effect, some cases, otherwise mysterious, can be thus explained.

You will meet with few cases of epilepsy in which the fits can be traced back to infancy which do not fall into one of these three classes, and you will seldom have any difficulty in discerning to which class a given case belongs. Of the small remainder, most are associated with congenital mental defect, and occur in families with neurotic disposition, and in which other instances of idiocy can be heard of. In these the malady must be ascribed to a congenital imperfection of the nerve tissue, of which the convulsions and the mental defect are both consequences. Such cases are usually self-evident. In the others, those which I have considered in this and the last lecture, the diagnosis is determined by the character of the fits and of the initial convulsions, and the circumstances under which these occur. But there is one important effect of the influence of attacks in disposing to their recurrence. After long continuance of convulsions, which begin locally and spread to the other side, the spread becomes facilitated by the residual disposition, and its influence on the whole brain may be such that minor attacks occur quite like those of idiopathic epilepsy. They are typical attacks of the form which is understood by the term *petit mal*. You must not, therefore, let the occurrence of these have weight against other evidence that the original cause was a local lesion.

A still more important question is connected with this residual disposition, and its action in augmenting the tendency to discharge of structures far away from the initial seat of the process. The question to which I refer, is that of the removal of the part from which the discharge proceeds, the extreme instability of which is the source of each convulsion. The operation seems at first sight a promising one, but the condition with which we have to deal differs much from the state in which there is some removable disease outside the nerve tissue, and acting upon it, such as depressed bone. Should the disease proceed from tissue which has recovered from partial damage, and which is in the vicinity of the original lesion, it is necessary to take away all that is around the latter in order to afford a reasonable prospect of the cessation of the fits. This involves, however, an increased loss of substance in the motor region, and an increased loss of function. There is at once increased loss of power, and although this lessens, it is permanently greater than before. It is, however, seldom sufficient to be put against the entire arrest of frequent and severe convulsions. Unfortunately, in a considerable proportion of the cases this arrest is not obtained. After an interval of freedom, which excites hope, the attacks return. This is not, indeed, surprising. The operation is only undertaken after long years of recurring fits. Every discharge has helped to fix the tendency in the structures, and in greater degree the nearer they are to the original lesion. The operation itself cannot but have, on the adjacent structures, some of the influence which the original lesion had on those near it, and this fresh influence will be exerted on structures already accustomed to discharge. How profound the effect on the whole brain is in these cases, is sufficiently shown by the occasional occurrence of the minor attacks already spoken of. When the prospect of failure is added to the definite risk of the operation, which involves at least some danger that life may be terminated in forty-eight hours by acute cerebritis, its value must have been already reduced almost to zero by the fits to make the proceeding justifiable.

Underlying all the phenomena of epilepsy, whatever its cause and whatever its features, there is one fact which it is important to recognise and to realise. Convulsions are far nearer normal action than their startling aspect suggests. In health, the nerve centres are always ready for the instant vigorous liberation of nerve force. A perception of danger induces in an animal, and often also in man, motor activity as intense as that of an epileptic fit. The nerve structures, by their nutritional state, hold, ever ready for release, the latent energy which excites the muscles, and this in what we consider vast amount. Considered from the dynamical point of view, it is no doubt trifling, not to be measured by its manifestation in the muscular contraction it excites. But the perfect readiness, which underlies its instant release in health, underlies also its instant liberation in disease, and thus it is less difficult for us to conceive the fact that the apparent causes of epilepsy are often such as seem to us inadequate.

I have no practical facts of prognosis or treatment to state to you, which result from the causal conditions I have described, except this. The cases in which we can trace an organic cause are less amenable to treatment than are those which are purely idiopathic. When the infantile convulsions of rickets are the remote cause of the disease, the prognosis is only influenced by the long duration of the malady. But when an old organic lesion gives rise to the affection, and is perpetuated by its extensive influence on the functional action of the motor structures, the treatment which is useful in the idiopathic affection is less effectual. Exceptions there are, and most cases are susceptible of benefit, but on the whole we can feel less hopeful of effecting great good in such cases than in others.

LECTURE XVI.
NEURALGIA.

To-day, gentlemen, I propose to ask your attention to a disease that you are certain to meet with, and certain to find among the most formidable, the most difficult, and even the most distressing of the practical problems with which you have to deal. The patient before you is suffering from senile neuralgia of the fifth nerve.

The mystery of pain presses upon our life on every side. It presses upon us in all our work, in every branch, but nowhere in such pure intensity and penetrating character does it tax our energy and baffle our resources as in that which is called the disease of pain, "neuralgia," "nerve-pain," because, in its special form, the pain is associated with some single nerve, as is that which is before us to-day. The prefix "nerve" refers to the limitation of the pain, and the name indicates the real character of the malady as a pure *disease of pain*.

The part which pain plays in life is varied, but for it we should be, on the whole, thankful. Pain is essentially a warning of that which is worse than itself. It is a warning which incites escape from the coming evil of which it is the shadow thrown before. Without pain no life could long endure. Without the warning which pain gives, various influences by which life would be ended could not be perceived in time for its preservation. The great office of the sensory nerves which the human and animal frames possess in such abundance is to give indication of that

---
Post-graduate Lecture, *International Medical Magazine*, March, 1895.

which is external to the organism. It is to permit the outer world to act upon the nervous system, and through the nervous system to enable the organism to act upon the outer world. Through the sensory nerves all knowledge comes; through the sensory nerves all guidance is afforded, and through them all warning of evil is obtained. The great agency of warning is pain. Without pain, without that which is incongruous, unpleasant in various degree, from the slightest discord to the intensity of agony, there could be no adequate escape from that which is injurious.

Yet it is scarcely correct to describe the influence of the external world as influence exerted through the *sensory* nerves. The accurate phrase would be that it is exerted through the *afferent* nerves, through the nerves which conduct impressions to the central organs and by the agency of the central organs exert adequate effects upon other structures. We are sensitive, sensitive enough. We have more than enough of the impressions which acting upon and through our nerves reach consciousness. We may be thankful that there are no more. But, as a fact, it is probable that the afferent impressions which act upon consciousness are not more than one-tenth of the total amount of impressions which act upon the centres. Consider for a moment, as I have more than once asked students to do, how much of which we are unconscious we may become conscious of. Consider at this moment how from any part of your frame you may become conscious of sensory impressions which before you were wholly unaware of. I have sometimes asked a class of students to fix attention upon the vertex of the skull and note if they did not observe a distinct sense of pressure there, and I have been invariably successful in the result. I confess, however, it has been at the end of my lecture, and not at the beginning. Therefore I am doubtful whether, at this moment, the effect will be sufficiently definite to justify my request. But the impressions of which we may become conscious by attention are but a small proportion of those which are forever passing to the central structures from every part of the frame, and the reality of this is shown by the fact that when intensified they are perceived as pain.

Consider the significance of the pain of pleurisy, of peritonitis, of inflammation of the serous membranes. There is agony most intense, spontaneous, and produced by the slightest mechanical stimulus. The path for those sensations must exist, and must be always a path of actual impressions, although they never reach our consciousness in normal states. This reserve, as it may be termed, of capacity for pain seems true of every structure, and everywhere we must conceive its significance to be the same,—that impressions in health are forever passing along the afferent nerves, impressions produced by varied processes of movement in the blood-vessels and perhaps the processes of nutrition, but impressions also which have their reflex influence in guiding the other subsequent processes (as from the muscles) related to those from which they result, but which never reach our consciousness. Yet there we have the faculty, the capacity for pain, as an ever-present potentiality in every structure and in every tissue, in general a safeguard against harm, harm of disease or harm of injury,— and in disease demanding, with a voice that cannot be unheard, the needed rest. Yet this, by morbid action, may be developed to a disaster. The excessive development of the higher functions of the nervous system, which comes with civilized life, must, it would seem, entail an excess of the guarding capacity for pain. Consider the influence, through generation after generation without number, of an existence which we cannot conceive as really natural, at least if we look at the life of the red Indians, or of other races whose habits conform to the general relations of their organism. Civilized life, that is, a mode of existence quite

FUNCTIONS OF THE FIFTH NERVE.

out of harmony with our animal frames, must entail disproportionate development and elaboration of the parts of the system on which it specially acts. It must, of necessity, develop this danger-signal of pain into a degree of intensity and readiness of excitation which makes it at last one of the grave evils of life.

Of that result we have the most conspicuous instance in neuralgia of the fifth nerve. This, surely, is natural. Consider what the fifth nerve is. I have alluded to pain as a warning, a mode of guardianship. Guardianship is needed in proportion to the preciousness of that which has to be preserved. The capacity for guardianship is proportioned to the readiness with which warning is produced, and to the effective degree which it readily attains. When we consider what the fifth nerve is and what it has in its charge, can we wonder that in this we have sensitiveness to pain raised to the utmost degree and developed to the potentiality of disease? Within its charge are the chief organs of special sense: the organ of sight, with all its marvellous delicacy of structure; the organ of hearing, hardly less elaborate; the organ of smell, with its relation to the entrance of air; and, lastly, not only is the organ of taste within the charge of the nerve, but the nerve-fibres of the fifth nerve, so far as its root is concerned, are actually the channel by which impressions of taste reach the brain. The nerve has thus in its care the chief avenues for knowledge, and it has also within its guardianship the avenues by which matter, and all the energy that matter bears, enter the system,—the opening of the alimentary canal and of the air-passages by which the oxygen passes to the system without which the food could have no influence. It is the guardian, as it were, of the gates through which enters that on which life depends and by which its power is supplied. It is, therefore, no matter of surprise that its functions should reach a perfection, a capacity, which render it above all other nerves prone to disorder.

In considering pain as a symptom of disease it is important to remember how much all our conceptions are colored, and how their form is determined, by our sensations. Consider the absolute distinction which we make between light and heat. We think of them as totally different things; and yet it is a fact that the undulations of light in the red end of the spectrum which are perceived by the eye as red light are absolutely the same as those which are felt by the skin as heat. It is not that the heat and light coexist at the red end of the spectrum: they are the same. We give them different names simply because we have structures that respond to them in different parts and of different characters. As our sensations seem to us so different, the causes of the sensations seem to us utterly unlike.

That which is true of kind is often true of degree. As we regard the same influence as different because its effects are different, so we regard influences as great or small as their effect on us is considerable or slight. There is no necessary proportion between the intensity of pain and the degree of the impulse which causes it. There seem to be special nerves through which pain is normally produced, which we call "nerves of common sensation." But there is evidence that by many other nerves pain may be caused, if the impulse which when in moderate degree gives rise to another sensation reaches an intense degree. In the nerves of "common sensation," the nerves of pain, a most intense sensation may be produced by a trifling influence. Conceive what may be taken perhaps as a minimum of nerve-impulse, the instance which I gave in a recently published address, the lightest touch of a hair upon the skin. Conceive the actual amount of energy constituted by the nerve-impulse to which that gives rise, which nevertheless passes up to the brain and produces its definite effect. As I have said, to conceive it you must attenuate your ideas of energy to atoms. In the nerves of pain it is probable that intense pain may be produced by a nerve-impulse scarcely greater in actual dynamical amount. It is important to realize this, because it enables you to understand why the morbid process to which neuralgia is due should so often escape detection, should, indeed, be beyond our present discernment. We have another illustration of it in the neuralgic pains of tabes. In a case of absolutely stationary tabes, with degenerative changes of the peripheral parts of the sensory nerves, there may occur from time to time, through years, paroxysms of agonizing intensity, and yet through all those years, with these attacks of pain coming and going, there is, as far as every other evidence can show, not the slightest increase in the disease. The changes in the extremities of the sensory nerves suffice to cause intense pain from the impulses which pass upward, impulses which may be an intensification of such as should be unperceived or may be produced *de novo*. How slight may be the difference in the character of the impulse which passes up unfelt or causes pain may be shown by this, that a mere change in the weather will cause an attack of intense pain.

As a matter of fact, in most cases of neuralgia of the fifth nerve, such as that which I am about to show you, no organic lesion has been found. It is a matter which is still uncertain. Facts can come but slowly to tell us whether most cases of persistent neuralgia of the fifth nerve are due to a peripheral or a central cause. It is most difficult to discern evidence one way or the other from the symptoms themselves. Among the facts that we have learned which seem of clear significance, there is the fact that disease of the centre of the fifth, in the pons, may give rise to such pain, that disease of the fibres in their course, perhaps also in the Gasserian ganglion, may give rise to such pain, and that disease of the peripheral termination of the fibres may give rise to such pain. It is important, in connection with the conception of the pathology of neuralgia, to realise the fact that the nerve-fibres which conduct differ only in degree from the structures which generate the nerve-impulses. I have elsewhere laid great stress upon this fact. It has been brought out by recent discoveries regarding the structure of the grey matter of the nervous system. These discoveries have an important bearing upon almost every problem of disease. The nerve-fibres, those, for instance, of the fifth nerve which reach in the pons and disappear from perception in the spongy substance of a long column of grey matter, probably end in that substance by terminations slightly thickened, it may be, but absolute. We used to say, we used to think, that the nerve-fibres end in the cells of the grey matter. The cogency of the facts which have been brought forward prevents me, for my own part, from entertaining any doubt of the correspondence of the new statements with fact. We must conceive that from the cells of the nucleus of the fifth nerve branching processes go off which end by contiguity, but not by continuity, with the ends of the fibres that conduct. We must conceive that the impulses which pass up leap in some way from one to the other. I say "leap," but it may be that the most powerful microscope would not reveal the chasm over which they leap. Yet it is absolute; a break in the molecular continuity which permits the passage of chemical action. The leap is probably by simple motion, like that which produced the impulse in the first instance at the periphery, where branching processes of the cells (that is, of the long fibres) end or begin in slightly enlarged terminations in the skin.

I spoke of "branching processes," but I should rather say "separated fibrils." I need not go into that distinction, although it is one of fundamental importance, because it is explained fully in an address I have lately published, and in which these facts and opinions are readily accessible to you.* Remember how extremely complex is the spongy structure of the grey substance in every nerve-nucleus; that we have innumerable fibrils in close relationship, and among these there must be terminations which constitute definite paths in the closest contiguity, so that impulses readily pass from the one to the other. Others must be sufficiently near for an energetic impulse to pass between them, but not for a slight impulse in the closest, and among these we can readily understand that there may be a difference in special excitability at the time. An impulse may even reach the centre and excite an adjacent fibril ending more readily than the ending which is in strict relation to it, because the former is the more excitable. If we conceive, as I think we must, that the nerve-impulse which begins by excitation from simple motion, which is propagated by chemical processes as a form of motion, may pass again as simple motion where the absence of continuity prevents the chemical transfer of its special form of motion, it is not difficult to understand that adjacent fibrils may be thus excited, and that we may have what is called an "irradiation of sensation."

I remember a personal instance of that. Unpleasant as personal experience may be, never forget that the most useful knowledge of your professional life is that which is subjective. I remember having a carious tooth in the lower jaw, from which I suffered no pain. After a time I had an attack of recurring paroxysms of intense neuralgic pain in the upper jaw, just opposite the tooth, but never associated with any pain in the carious tooth. At last I

---

* "The Dynamics of Life." London: Churchills; Philadelphia: Blakistons.

had that tooth extracted. The process of extraction caused the most intense paroxysm of pain in the upper jaw that I had ever experienced, and—it was the last. There can be no doubt that the irritation of the inferior dental branch of the fifth nerve had in some way, probably by some increased susceptibility of contiguous fibrils, led to their special excitation, and their stimulation had reached a morbid degree, intensifying itself until it possessed the capacity for neuralgia.

These new facts of the relation of the nerve-structures do enable us to understand many phenomena better; and that which explains much is seldom wrong. We can sometimes discern truth most surely by perceiving how much that is obscure is made clear, how much that is discordant is brought into harmony, how much more we can look for than before could be conceived.

Among the many varieties of neuralgia which you will find arrayed in text-books in serried ranks of formidable length and imposing order, three groups are specially important: (1) The neuralgic pains which occur at all periods of life from definite local irritation, such as that which decayed teeth induce. (2) The changing neuralgic pains that we meet with especially in middle life,—when a neuralgic pain comes in one part, then passes to another, then may leave and go to a third, then perhaps will change to some other temporary neurosis, and then may vanish as an altogether different malady comes on. (3) The class of which this case is an example,—senile neuralgia.

Neuralgia in its intense severity and enduring form is a malady of late life. It is the most formidable, the most distressing disease that life can bring, and it comes when the clouds should clear for the placid sunset, always longed for, seldom obtained. It is a disease of age. The influence of age on neuralgic pain is an important and significant fact, shown in a most striking way by the neuralgia which follows herpes. After the age of fifty years post-herpetic neuralgia is prolonged in proportion to age, or rather its persistence is longer and far greater than the ratio to age. In middle age the neuralgia is trifling, and will last only for a week or two, but at seventy it will last for years, and may never pass away entirely. That is true irrespective of situation. I remember that Sir William Jenner used to tell us of one striking instance of this. A man had herpes zoster on the calf of the leg. It was characteristic, and had left the usual sequel. It was before the days of chloroform; he had everything done which could be thought of, and at last he consented to endure the pain of having that part of the skin and muscle cut away. Obtaining no relief, he shot himself. That may impress on you the intensity of the pain in the old which is trivial in the young. Why it should be I do not know, but we may form some dim conjecture when we remember that all these nerve-impulses are matters of, as far as we can see, chemical processes occurring in the tissues under the influence of life. The vital power of nutrition is the influence which renews the capacity for function, by replacing the molecules which have been changed by the functional action which has just occurred. As life goes on, the capacity for renewal becomes less perfect, so that molecules are formed less competent to achieve their purpose, more prone to give rise to abnormal impulses, and recovery from any other morbid process produced by outside influences is less perfect. The restoration of structure is not such as to enable it to do the normal work in perfect degree, and the degree of imperfection of constitution, so slight, it may be, as to us to be scarcely conceivable, may determine a morbid function adequate to produce intense pain where under normal condition hardly any sensation should be caused. Every morbid functional action is followed by a renewal of capacity for like action, but renewing it every time in an increased degree. The nutritional power of life is an augmenting influence, potent for all acquisition of capacity, healthy and diseased. The disorders of the nervous system that depend on morbid action are, by the vital processes of nutrition, self-perpetuating.

This patient is forty-seven years of age. Her case makes us call to mind as possibly important the fact that senility is individual, so far as concerns the time of life at which age becomes "old age," and come the transition must, unless the shears snap the thread before it breaks. It is not only individual, but often partial. Every grey or hairless head reminds us of the fact, and should prepare us to meet with local troubles in some patients which are commonly met with only in those who are much older. The patient is younger than most subjects of her malady. But in life's sad cadence, tones only just distinct in one generation dominate that of the next. The patient inherits neuralgia, but not in its senile form. Her mother suffered from nerve-pain for several years, but in another seat and form. The transmission of tendencies to disease rather than of precise disease is forever rising conspicuously before us when we endeavour to discern the relations of the morbid processes we have to treat. Each tendency may be inherited in hindered or augmented form, often through causes that we cannot trace. Senile changes in one person may come only a little before due time, and in the offspring may be definitely premature. The strength of tissue vitality varies, and is capable of variation through direct and indirect influences. Its primary influence is on function, the changes in nutrition first thus manifesting their presence. Ultimately derangement increases to loss as nutritional alteration advances to structural change. But this is clearly seen only when function readily reveals the early change.

There is another class of cases in which the inheritance

of disease depends upon the inheritance of structural peculiarities, unimportant in themselves, incapable of manifesting their presence, and yet determining in an adventitious manner grave disease. Thus early death may be inherited through that which, save for its position, would entail no difference in any one of life's many features. Yet a simple inherited peculiarity in anatomical arrangement may entail inherited disease. I remember a peculiarity in the retinal artery that was the same in mother and daughter. I cannot doubt that a like inheritance of arterial branching may determine strain on certain cerebral vessels, and early atheroma so situated as to entail occlusion of an important branch. So also we may conceive that in nerves inherited characters may determine definite effect, and that far more frequently, and far more powerfully, inherited tendencies to nutritional failure or nutritional susceptibility may induce functional disorder. It may be transient in middle life, but enduring when in later life the vital power fails to maintain adequate nutrition,—that is, the due appropriation to the molecules of the elements on which the proper action depends. We can understand that the liability to derangement at one age should be inherited as a liability to later failure, and that thus this woman should now present the most intense form of senile neuralgia, whose mother suffered from varying though severe neuralgia in other parts at an earlier time of life. It is four years since the pain, now so intense as to make her life one long agony, began. A decayed tooth was thought to be the cause of the pain. Many such irritative causes of neuralgia are real, although their removal may not cure, as with tapeworm and epilepsy.

Epilepsy—*i.e.*, repeated convulsions—may be due to tapeworm. It may be expelled, with its head, once and forever, but the fits go on, just as all the consequences of evil deeds live afterwards.

Paroxysmal pain is the analogue of paroxysmal spasm.

It is facilitated by repetition, as is every functional process by the residual effect of augmented capacity. In this patient, moreover, some constitutional condition of peculiar character may have been at work. An inherited disposition may be augmented by another. The patient had, during each pregnancy, neuralgic pains in one cheek. The influence of pregnancy is often definite, but seldom the same. This morning I saw a lady subject to intense migraine who never had an attack during pregnancy. The relation of the state of pregnancy to the nervous system is alike important and mysterious. Why it should induce or hinder the same morbid state we do not know. We shall learn some day, and with it much more. The last attack in this patient began eight months ago, and has continued. It was sufficiently severe to cause the extraction of all her teeth. The pain persisted though the teeth were removed. Above the upper molars is, as you well know, a curious cavity called the antrum, a convenient receptacle for conceptions of disease. Her antrum was then opened, apparently with the object of permitting the escape of whatever disease it might contain. Whatever was released, the pain remained. The operation did no good. But morbid processes do not always avail themselves of the opportunities offered them. So then her antrum was drained. But the source of pain was not even drained away. Yet some effect resulted. She improved slowly. We should remember that many remedial measures which do not do good directly do harm indirectly by lessening the patient's strength. This is a rule applicable to many diseases, which is rather apt occasionally, in the flush of therapeutic energy, to be lost sight of.

She is now suffering from many attacks every day, or was until she came here, and each begins by a curious sensation, which is not actual pain, near the junction of the left lower alveolar margin of the ramus of the jaw. She describes a flickering sensation at the extremity of the upper gum. This soon becomes a darting pain, which spreads to the lower jaw and seems to pass upwards and backwards. Sometimes it begins in the lower jaw. Any movement, any touch, will bring on a paroxysm, especially if it is a little time since she has had one. The attack seems in some way to exhaust the tendency, so that a touch immediately after an attack will not cause one, while the same touch would cause one after an interval. It is just as with migraine: a patient who, a month after an attack of migraine, could not eat a mince-pie without another attack, may eat two such evil but pleasant things a few days afterwards without effect. So also it is with epileptic attacks. The recurring influence of each functional process is to induce a nutritional development of renewed tendency, which takes some time to reach a high degree. Except just after an attack, any movement of her tongue in that part of the mouth will excite a paroxysm, but the more sudden the movement the more severe and immediate is the pain. I can show you the difference. You will see that a very slight touch, if sudden, will cause an attack of pain, but a very much more severe pressure may be made very gradually without pain. It is one instance of a general law. A galvanic current gradually increasing may be passed through a nerve without exciting it, but if suddenly increased it at once causes stimulation. This is true also of the interruption of the current. I can show you a similar effect in the mechanical effect of pressure on the sensitive nerves of the patient. I press very gradually until at last I am pressing firmly, and yet the patient says that it does not produce pain. I press suddenly, and you see at once evidence of suffering. I again press gradually, and gradually lessen the pressure, and there is no pain. Again I press gradually, and suddenly withdraw the finger, and pain is acute. The patient has had fewer attacks of pain during the past week, but she has had more slight continuous pain. The substitution of this for the intense paroxysms is probably an improvement, but the patient seldom realises it. A present evil, vivid in its experience, always seems greater that that which memory only presents to the consciousness. The slighter constant pain seems to her harder to endure than even the frightful paroxysms she suffered with intervals between of perfect freedom. The patient has been admitted with a view to an operation upon the nerve, but at present we have only an example of the frequent difficulty that is due to some improvement produced by the perfect rest or by the treatment it is right to try.

The result of division of the nerve is sometimes great and sometimes most disappointing. Benefit apparently depends upon the effect of preventing all afferent impulses from the periphery. When the neuralgia depends upon nutritional changes in the centre, rendering it unduly active, so as to generate impulses, the activity may be kept up, and the paroxysms of pain caused by impressions from the periphery which normally would have no effect. If the morbid process is actually in the periphery, or in the nerve-fibres which conduct and are also excitable, and the nerve can be divided above the morbid process, then the morbid impressions are arrested. But it may be that even then the effect is not absolute, because the repeated abnormal impulses giving rise to the pain may have brought the centre into a state of spontaneous over-action, and neuralgia, primarily peripheral, may be ultimately central, and thus continue when the peripheral cause is removed from action. It is a similar process to that of epilepsy due to a tapeworm in the intestine, and may persist for years after the cause is expelled. Such arrest of the peripheral impressions is a point of great importance in treatment, but it is so formidable that only when other measures have failed can it be proposed. In many cases, especially in those which are not of long duration, we have the means of lessening the peripheral impressions for a little time each day by the hypodermic injection of cocaine, and, slight as this may seem to be, yet by its repetition it has often in time an unquestionable influence. Moreover, the importance of combining several influences, each of which alone has but slight effect, is often strongly impressed on the practitioner. It interferes seriously with the progress of therapeutic science, but we have to gather for the general good the fragments that we can secure without individual harm.

I can glance at some only of the more important of the other general elements in treatment. Apart from the removal of causes and the operative treatment, the chief measure is the promotion of the general health by every means in our power, avoiding whatever lowers the tone of the nervous system, and endeavoring to strengthen it by drugs, and to lessen, if we can, the tendency to its over-action in the same way. I do not hesitate to speak of treatment by drugs. I hold that their influence is by conveying energy in special forms to special structures in which energy is evolved. I wish we could associate drug with drag and draw. Alas! Professor Skeat forbids, and from his decision there is no appeal. He only permits us, as the source of the word, dried roots and sugar-plums. Well, "these things are an allegory," without doubt, but not that which I desire. In passing, may I ask if you know Skeat's "Concise Etymological Dictionary," in which words are arranged according to their derivation? If not, let me urge you to spend on it the first seven shillings and sixpence you can save or beg (I will not go farther, though I almost might). But, frankly, if a copy cost five pounds, I would sell any books I have, to that value, in order to obtain it.

To see the real significance of the use of drugs—as dynamical therapeutics—we must realise that all the impulses in the nervous system are the result of chemical processes occurring in the nerve-tissue. In this the molecules, their composition and arrangement, are determined by the influence of life, mysterious, inscrutable to us, perhaps, forever. But the processes depend on energy latent in the molecules, released by chemical union. Most drugs, like most food, are of value, not as matter, but because they convey energy. These chemical compounds present latent chemical energy in certain forms, and by entering into the composition of various structures they modify their composition or action or both: thus they do good; thus also at times they do harm. But into this question, fascinating as it is, I cannot go. You will find its grounds discussed in the address I have already referred to.

In the treatment of neuralgia I confess my own experience does not lead me to express any high estimate of the older drugs, such as sulphate of copper, but the influence of direct sedatives is unquestionable. The influence of cocaine is purely local—it is simply to arrest the afferent impulse; but the influence of morphine is central—it has special action upon the sensory structures. A strange thing, which may not have struck you, is that morphine and opium seem to have a special action upon the centre which is over-acting, so that the agents will quell pain without producing sleep, and indeed at first in doses too small to cause sleep, if only they reach the centres with the sudden momentum secured by hypodermic injections. As a rule, in my observation the greatest benefit has been obtained from the milder sedatives, especially Indian hemp and gelsemium, while the tonic effect that is usually essential for permanent benefit is produced, I think, even on the sensory structures, by strychnine more effectually than by any other drug. Its effect also seems to be proportioned to the momentum with which it is brought to bear upon the elements, and as it is not always convenient to give it as a hypodermic injection, and it is well to convey the momentum of its tonic influence with the momentum of gentle sedative influence, I have been accustomed to combine strychnine with Indian hemp or gelsemium, and to secure their more rapid transit by giving at the same time nitroglycerin. Our pharmacopœial one-per-cent. alcoholic solution called tinctura trinitrini is most convenient for the purpose. The important thing is that the initial dose should be uselessly small and rapidly increased, and that the mixture should not be alkaline. From this I have seen results, even in such cases as the patient before you, exceeding anything I could have expected, giving greater and more permanent relief than any other therapeutic measures, so much so that the treatment has made it unnecessary to operate on some patients who were admitted for that express purpose.

LECTURE XVII.

LEAD PALSY.

Gentlemen:—You remember the case of lead palsy, or wrist-drop, which I lately showed you. I wish to-day first to mention some practical points in connection with the affection, and then to point out to you the general aspect of the disease as it is revealed to us more clearly by the recent increase in our knowledge of such causes.

The first practical point relates to diagnosis. The palsy is so generally connected with lead, that it is seldom mistaken. The danger is lest other things should be mistaken for it. Yet the practitioner or student at once looks at the gums, and the lead line at once indicates to him that the cause which he expects is present. But it is important to be aware of the fact that this trust is only justified when the gums are not perfectly close to the teeth. A like palsy is the result of other poisons; but in their case there is to be found no line upon the gums. Alcohol, arsenic, and perhaps other metallic poisons may cause the same palsy, but the absence of a line upon the gums excludes lead as a cause on that one condition—provided the gums are anywhere separated from the teeth by a space in which there are decomposing albuminous materials capable of yielding sulphur. With perfect gums you can only exclude lead-poisoning by excluding its possible causes, especially by having the drinking water analysed, and by having the urine analysed after iodide of potassium has been taken for a week.

Clinical Journal, May 1, 1895.

But the most difficult point in diagnosis, in my own experience, is due to this strange fact, that a chronic spinal muscular atrophy occasionally begins with rapid weakness of the extensors of the arms on both sides, having exactly the onset and the localisation of lead palsy. In the cases I refer to, of which I have seen several, this has been followed by spreading muscular atrophy elsewhere in the arms and trunk, and by other indications of morbid process in the spinal cord. In none of the cases was there a verification. Yet I could entertain no doubt of the truth of the fact, especially as every toxic cause was excluded. In the early stage, when there was simply the rapid weakness, with degenerative reaction in the extensors of the forearms, precisely localised as that of lead poisoning, the diagnostic question was one of the most difficult that can be conceived. As soon as spreading atrophy was obvious elsewhere, the question became simple; but it has happened that the difficulty was, at first, the greater in some cases I have seen, because the gums were so perfect that the absence of a lead-line was unimportant. Its absence had none of the significant meaning its presence would have afforded. In such cases you have only to wait and watch. In other cases of toxic origin you have, as a rule, sufficient evidence of the poison, from other symptoms, or there is weakness in the corresponding muscles of the legs which you know to be such a characteristic of alcoholic palsy, and is scarcely ever present in lead palsy. That is also the case with arsenic. Wrist-drop may result from arsenic precisely like the wrist-drop from lead, but there is also similar palsy in the homologous muscles of the lower leg—those which flex the ankle and extend the toes. I think I have mentioned to you an instance of that which came under my notice—a case of symmetrical palsy of the arms and legs in a lady who had been working for years on some muslins. They had been bought as possessing a specially "æsthetic" colour. But a "thing of beauty" is not always a joy forever. The muslins were found to be charged with arsenic. The palsy lasted long, but recovery was ultimately completed.

There is an interesting question which may sometimes be important to you with regard to the "lead line." You know that the treatment of lead-poisoning consists in giving iodide of potassium. And why? Iodide of lead is absolutely insoluble; why, then, do you give iodide of potassium? For this reason,—and remember it as an instructive instance of the necessity for caution in letting theory determine practice,—that, although iodide of lead is absolutely insoluble, by giving iodide of potassium you increase the quantity of lead in the urine. Here, then, is a practical fact on one side, and a theoretical fact on the other. I yield to no one in regard for theory, but theory is nowhere beside fact, and you do eliminate the lead by giving iodide of potassium. The fact is, that lead is in the system in combination with albuminous substances, and the iodide seems to form a combination with them that is soluble. The effect, therefore, of giving the iodide is to increase the excretion of lead, and, of course, to do this it has to increase the amount in the blood. You have, therefore, to be a little careful, when much lead has been recently taken in. Before you give the iodide, you should clear out the lead from the alimentary canal by giving an aperient for a few days; and then you must give but little iodide at first. That is the essential treatment.

But the point I was going to mention is, Does the iodide remove the lead line from the gums? The lead is in the form of sulphide of lead. We cannot *a priori* infer that the iodide will act on the sulphide of lead as it does on the lead in combination with organic substances. My own observation has afforded me no instance in which the iodide of potassium has removed the lead line; and I have seen cases in which, after the ingestion of lead had ceased for some years, the lead line was as perfect as ever, although iodide of potassium had been taken at the beginning for many months. This is not a theoretical question only.

You may be in doubt whether given symptoms are due to lead in a patient who, say, a year before had certainly been exposed to the risk of lead poisoning, or had had lead poisoning, had been treated with iodide of potassium, and had gums such as would give rise to a lead line. Is the absence of the lead line of significance as regards the question whether present symptoms may be still due to lead? May the lead line have been removed, and some lead be still elsewhere in the system? I think not. I think that if lead still existed in the system there would certainly be a lead line. The point may come under your observation practically in more than one way. A case which brought it very prominently before my mind was that of a middle-aged man who had been gradually losing his sight with symptoms of chronic slight optic neuritis, first observed about a year before, and still going on. For a long time he had been sufficiently anxious to increase the growth of hair upon his head as to induce him to use a largely-advertised preparation, which on analysis was found to contain lead. The man brought an action for damages for his blindness, and I was consulted. There was no lead line, and yet his gums were such as to afford every facility for its formation. I felt sure from the absence of the lead line, and the fact that the optic neuritis was going on still, although it was more than a year since he had discontinued the use of the preparation, that the condition could not be due to this cause. The proprietors of the substance, however, afraid of the damage which the mere accusation would entail, compromised the matter. This case is an illustration of that which may happen to some of you, and the importance of the point. Many practitioners have ample opportunities of observing all the effects of lead poisoning and of making very important observations; such opportunities may come to some of you, and should not be lost.

A point in treatment to which I have not alluded is the use of electricity. That which it can effect is, as in all cases of disease of the nerves, to help in maintaining the nutrition of the muscles while the nerves are recovering. That it can do this to some degree I have no doubt. I have seen some clear instances of its service. Before the nerves recover, so as to conduct the motor impulses to the muscles, the condition of these may be distinctly improved, their response to the will promoted, and its degree increased by the application of voltaic electricity. I remember a gentleman who suffered from wrist-drop, and whose wife suffered from neuralgia. In neither of them was there a trace of lead line, and the cause of the affection had therefore been entirely overlooked by at least one very eminent physician, the fact that the gums were perfect not having been taken into consideration; but the cause was suspected by another physician, and it was found that the drinking water was charged with lead. From the failure to recognise the cause, absolute palsy had gone on in this case for nearly a year. There was no power whatever in the extensors, and on the first application of electricity not the slightest response could be obtained. After several applications, however, the muscles began to respond to the stimulation, and continued application led to a gradual increase in the degree of contraction that was excited, until, in the course of a fortnight, they contracted well. Soon afterwards slight voluntary power of contraction began to return, and it steadily increased, and the patient made ultimately a perfectly good recovery. In such a case, in which electricity so clearly abolished the absolute inertia, there can, I think, be hardly any doubt that the local stimulation which voltaic electricity affords to the muscles is of definite service, and possibly without it, the voluntary impulse would never again have excited them.

Lead may also cause another form of palsy, not affecting specially those muscles, but affecting first the small muscles of the hand—there is, in this form, gradual wasting and gradual weakness, not, as in the extensors, rapid weakness followed by wasting. It is a condition identical, in its local symptoms, with that of progressive muscular atrophy, due to disease of the nerve cells in the cord, and probably identical in its seat. It seems to depend on a gradual wasting of the cells, fibres, and muscles under the influence of lead. In this form, as in progressive muscular atrophy, there is merely a diminution of the irritability to both forms of electricity, equal, steadily increasing from the first, there being no period of lowered or lost faradic and preserved voltaic excitability. It may in rare cases be wide-spread. Its recognition is of great importance, because it does not improve and recover as the wrist-drop does. The wrist-drop may pass away entirely and this slow wasting away remain unchanged. If the ingestion of lead is stopped it does not as a rule increase, but I have not known it to pass away.

Lead palsy in its common form is due to an affection of the motor nerves; the motor nerves to the muscles that are affected. The acute atrophic palsy always shows that there is an acute affection of the nerve fibres, or of the nerve cells from which they spring. Whenever there is any acute disease of the nerve fibres, their endings, or of their special cells, there is loss of faradic irritability in the muscles, and preservation of the voltaic, and there is loss of both in the nerves. The electrical reactions do not show whether the disease is in the nerves or in the nerve cells of the cord; and until within the last few years it was uncertain to which the wrist-drop was due. It has been proved to be due to the nerves, and it has been proved to be only one of a series of similar palsies which are due to changes in the nerves and are produced by a toxic agent. Before we had discerned the characteristics of these remarkable toxic palsies, and the evidence that they depend on the influence of the poison on the nerves, our uncertainty regarding its seat was increased by another circumstance, the occasional correspondence in distribution of central and peripheral affections. I have already said that cases of spinal progressive muscular atrophy sometimes begin by rapid weakness of the extensors of the wrist, quite like lead palsy, with the same localisation, even to the escape of the supinators and of the flexor of the metacarpal bone of the thumb. What does this show? It shows that there is a similar pathological susceptibility on the part of the motor nerves and on the part of the cells from which they come. This is a fact of the most profound significance. It shows that the peripheral structures and the related central structures suffer in the same association. In each corresponding structures have the same susceptibility, and seem to have this same associated susceptibility to different morbid influences. We can thus understand many phenomena otherwise difficult to comprehend.

Another point regarding the lesion in lead palsy deserves attention. Lead palsy is a painless affection, and yet there is an acute change in the nerves, a change which consists in that rapid alteration of the nerve elements that we call parenchymatous inflammation—inflammation of the special structure. It is associated with some indications of inflammation in the sheath and other connective tissue elements of the nerve. You know, however, what an intensely painful affection neuritis commonly is. Yet here we have a neuritis which is painless. What special feature can we perceive to explain the difference? The nerve branches affected are those to the muscles; noting that we find other illustrations of the same fact. We meet occasionally with an isolated one-sided neuritis of a motor nerve; a motor nerve, for instance, of some of the calf muscles, which is absolutely painless. The absence of pain, as I have known, is most liable to mislead in diagnosis. But remember how essentially painless is neuritis of the great motor nerve of the head, the facial nerve. There may be pain from associated morbid states, due to the cold that causes the neuritis, but there is no pain from the inflammation of the nerve within the Fallopian canal. It would seem that the sheaths of the purely motor nerves and these branches do not possess such sensory fibres as most nerve sheaths do, and, therefore, their inflammation is not a source of pain.

Lead palsy, then, is an example of an affection of the peripheral nerves from the action of a poison. But it is an affection of a few nerve fibres among many of like motor function, an affection that is symmetrical, similar on the two sides. This is the important fact to be realised by every one as the great indication of the whole class of disease to which I refer; the limitation of its functional incidence, and with this its bilateral symmetry. These two symptoms indicate one or two processes, and in many cases they indicate both. The two processes are degeneration and the action of a poison. We are beginning to see in how very large a number of instances a process of degeneration is the effect of a poison. You will find, if you keep that law clearly in mind, that it will guide you right in innumerable instances of the most diverse aspect, in which all sorts of motor and sensory and reflex symptoms, in various places, baffle your efforts to discern their cause, unless you note this character and know its meaning.

What does the symmetry mean? Symmetry indicates an influence reaching both sides of the body; and if it acts on both sides of the body it must be an influence which reaches all parts, because we are unable to conceive an influence which acts equally upon certain local parts on the two sides and cannot reach others. The bilateral symmetry so distinct in the features and the limbs is not a matter of the surface, nor is it a matter of the visible form of the surface alone. It obtains and must obtain in every structure, down to the minute differences of chemical composition which underlie function. Parts of the two sides that have similar relations and the same functions have, therefore, corresponding constitution, and it is that which is the cause of the bilateral symmetry of the effects of any morbid influence which reaches both. Consider for a moment how belladonna taken by the mouth acts upon both eyes, and how perfect that influence may be in limitation and symmetry. Consider the minute correspondence of the two sides, and the minute difference from other structures on each side, which this effect signifies. A similar condition alone explains the incidence of lead and other toxic agents. You can conceive of no structural relation or relation to vessels which determines why lead acts upon the branches of the musculo-spiral nerve, and not upon any other, and why it acts only on some of the branches. It can alone be due to some slight inherent difference in the minute constitution of the nerves which enables the atoms of lead to join with the atoms and molecules of these nerves, and derange the process of their nutrition acting on these nerve fibres and not on others. And, moreover, a like condition is the only conceivable explanation of the fact that lead sometimes, and other poisons often, act also on the fibres in the legs that are homologous with these fibres in the arms, and not on others. When we think of the fact, how great is its significance. Those nerve molecules, for some reason, have not the power of resistance to a morbid chemical agent that other nerves have. They differ in their behaviour when a molecule of a metallic poison is presented to them, although they are the same in the minutest scrutiny we can give them, the same to every other test we can apply. Whether any explanation will be found in future in some developmental process connected with the appearance of these muscles and nerves in the range of the animal kingdom, or not, I do not know. But the fact is certain that to all these toxic agents these are specially susceptible, and the fact is certain that it must be due to a greater readiness for the reception of the toxic agent and for its influence.

The various effects of different poisons bring before us, in almost startling clearness, how profound is the ultimate influence of differences far beyond our power of scrutiny. They show us how much of disease that is mysterious may be due, in this manner, to chemical agents, and they afford us a rational basis for the old faith, not yet lost, perhaps, indeed, reviving, in chemical agents as a means of influencing the processes of disease.

LECTURE XVIII.

SATURNINE TABES.

Gentlemen:—I wish to direct your attention to-day to a case that illustrates a subject to which I have repeatedly referred during the last few months. I should, indeed, hesitate again to refer to it were I not impressed with the importance of considering principles and laws, again and again, in fresh relations. You will remember that I have had much to say regarding the influence of poisons—of toxic agents—on the nervous system, and that I have impressed on you the fact that this is one of the directions in which medicine has made the greatest strides during the last twelve years. You know in how strange a way different poisons act on different parts. Strychnine disturbs the function of one part, atropine of another, arsenic of a third. You know, also, that this must be an indication of some otherwise unrevealed peculiarity in the structures acted on. If we observe carefully the effect of the poison we see that there is also a considerable difference, which depends upon the intensity of the poison. A poison which acts suddenly and in large quantity produces effects which are much less localised than are those of one that acts slowly and gradually. Many parts are susceptible to a powerful influence, with much momentum, though only a few may be susceptible to the slow influence of smaller doses. Intense poisoning by lead may cause wide-spread

---

Delivered February 13, 1895. *Medical Magazine*, April, 1895.

symptoms, including acute disturbance of the brain; while a gradual, slow poisoning may affect only the extensors of the wrists. In diphtheritic palsy we have usually an acute toxic process, due to the sudden production of a large amount of poison, which acts very widely and very intensely. But in other cases we may have a slower action on some of the limbs. One patient will have universal poisoning and muscular atrophy, with signs of nerve degeneration. Another patient will have only symptoms that present a perfect semblance of locomotor ataxy. The important fact that I wish you to realise to-day is that the poisons in their slower actions influence the parts most susceptible to them.

Yet to this law we have to recognise some apparent exceptions. I especially wish to emphasise the fact of exceptions, and the probable cause of their occurrence. A poison which usually produces one form of palsy, may in another case produce a different form, and yet we may be able to perceive no difference either in the amount, or in the rapidity with which the poison enters the system. The explanation is, I feel sure, that the precise form in which the poison is presented to the nerve elements is subject to variations. Even the metallic poisons are probably never presented to the nerve elements in any simple form. All these metallic substances seem to form compounds with albuminous bodies, and to be thus presented to the tissues. The compound formed depends partly on the dose, partly on the form, and very much, probably, on the precise composition and character of the albuminous material, which varies in the same individual according to the state of health and according to any constitutional peculiarity. Hence it is easy to conceive that in a patient with syphilis, or in a patient with a gouty constitution, the same metallic poison may be presented to the nerve elements in a different combination, and may be thus more or less influential on function and disturbing to molecular structure.

Why are these compounds so influential? Because every part of the nervous system and of every other system (though we are now only concerned with the nervous system) is in a state of constant functional action, and therefore in constant molecular change. Consider how constant is the functional activity of the sensory and motor structures. We are conscious only of a minute proportion of the impressions that are ever traversing even the afferent nerves; while from the motor centres efferent impulses are always acting on the muscles, maintaining their tone. This fact is true of all parts of the nervous system. Continuous activity is the universal law.

All functional activity is a matter of chemical change. The elementary molecules of nerve structure, formed under the influence of Life by the processes of nutrition, are disturbed in their action, and yield some of their constituents to the oxygen which is forever lying in wait. They are removed, and are instantly replaced, under the same power of vital nutrition, from the materials which the adjacent plasma has brought to them. And, therefore, if the plasma is abnormal in its constitution from the presence of a toxic substance, there is constant opportunity for that toxic substance to enter into the constitution of the nerve elements. However minute the amount may be, no minuteness can prevent the results of the persistent appropriation of the abnormal molecules. Phosphorus is a normal constituent of nerve substance, and you know how close is the relation of phosphorus and arsenic. So we can easily understand that when arsenic is circulating in the blood it may take the place, to some extent, of phosphorus, may enter into the nerve elements, may derange function and derange nutrition, and, last of all, may derange structure to visible degree.

When the influence of toxic agents acting slowly is considered, we find that in all cases it is chiefly on the nerve structures furthest removed from the centre on which their nutrition depends. I explained to you a short time ago how we have come to discern that the great function of the nerve cells seems to be the nutrition of the nerve fibres. The longer the fibre the greater is the distance through which that nutritional influence has to be exerted; and so the further the distance from the cell, the feebler this influence will be, and the more prone will be the fibre to suffer change of nutrition from any cause—the less able will it be to resist and reject an abnormal constituent, and the more prone to suffer gravely from it. This is probably the explanation of the fact, with which you are all familiar, that the common effect of these poisons, when they act slowly, is to cause degeneration of the peripheral parts of the nerves. A few, like strychnine, seem to act on the grey matter, but the majority, when in moderate quantity, act upon the nerve endings, and in proportion as their influence is greater its effect extends further and further up the nerve, although it seldom reaches the nerve cell on which the vitality of the nerve depends.

You know the cells in the cortex of the brain, from which proceed the pyramidal fibres to the spinal cord. These probably end by ramifications related to those of the motor cells, whence proceed similar processes which terminate upon the muscles. From between the muscular fibres, and also from the skin, we have afferent nerves passing to the ganglia on the posterior roots, and from these proceed the fibres that continue the tract, which pass—some up the cord and some to the cord at the same level.

As I have just said, if we compare the actions of the various poisons, we trace in most a tendency to act on the extremities of the fibres, the parts that are most distant from the nerve cells. For instance, the pyramidal fibre proceeding from the nerve cell in the cortex seems to be especially influenced by that peculiar poison which is contained in the seeds of leguminous plants, the effect of which is known as "lathyrism," and which causes symptoms of lateral sclerosis. Lead acts upon the extremities of the motor fibres in certain muscles; while the peculiar chemical poison which we are compelled to assume that syphilitic organisms engender seems to act on the parts of the fibres of the afferent muscle nerves, and also on the long fibres which ascend the cord from the muscles, chiefly in the posterior medium column, and seem to subserve cerebellar co-ordination. Generally it acts also on the sensory fibres from the skin. Alcohol may have the same effects, but prefers the motor nerves for the extensors. This is true also of arsenic. Both these usually influence the nerves of all four limbs,—influence them equally or nearly so, both motor and sensory. As I have said, we have curious anomalies in these effects. Lead commonly acts only on the motor fibres of the arms. You are familiar with the fact just mentioned that the poison we assume to be left by syphilis—I say "assume," but all analogy makes the assumption as near an inference as such reasoning can—that it has a special tendency to act upon the muscle afferent nerves. By this it gives rise to the loss of the knee-jerk and to ataxy, of which impairment of these nerves is the great cause. Although arsenic and alcohol commonly affect motor nerves, each may also affect the afferent nerves, including the sensory nerves from the skin, and sometimes, for reasons which we do not know, they affect especially the afferent nerves from the muscles, and give rise to symptoms like those of common tabes. Indeed, there is more than likeness, there is *identity*—identity of symptoms with the slighter cases of true tabes and identity of process as far as can be discerned. The alcoholic pseudo-tabes, with which you are acquainted, is an affection of the afferent nerves from the muscles, probably quite like that of some cases of tabes after syphilis. We may have a like condition from arsenic. For some reason the poison reaches the nervous system in such a form, and the nervous system may be so disposed, that the agent acts upon these nerves rather than upon any other. Instead of affecting the motor nerves, it acts on the sensory fibres, and instead of acting on all, it may act chiefly on the fibres from the muscles. Lead, however, affects the motor nerves almost exclusively —the fact is familiar. Yet there are cases in which it also causes loss of sensation. I have recorded a case in which lead caused loss of sensation in the skin around the anus, apparently from a peculiar action upon the peripheral sensory nerves.

Remember this, however, that most agents which act on the afferent nerves, influence, in most cases, in greatest degree those from the muscles. The affection of these nerves being the cause of ataxy, we should expect, from all these facts, that a form of pseudo-tabes would sometimes be produced by lead as it is from alcohol. It has been said to occur, but I have never seen an instance, until the patient whom I am glad to be able to show you to-day. The reason why I have introduced the case to you by this roundabout dissertation is in order that you may be able to see it without surprise and yet with interest. The action which these poisons generally exert, the variations which they present in their effect, and in the precise nerves on which they act, are such as to make it no matter of surprise that we should meet with a case in which the afferent fibres from the muscles have been chiefly affected by lead.

You see that the patient walks with his legs wide apart, and keeps his eyes fixed on the ground. When he stands still and puts his feet together, toes and heels, and shuts his eyes, there is far less than normal power of maintaining the balance. He cannot, indeed, keep upright. Observe his gait more closely as he walks. You see that there is considerable difficulty in maintaining his equilibrium, but there is not the peculiar irregular movement of the feet you so often see in ordinary locomotor ataxy. The precise symptoms of ataxy depend upon the muscles that are chiefly concerned, on what the muscles are, from which the upward impulses to guide co-ordination are deficient. When the muscles whose impulses are deficient, the contraction of which has not the proper guiding influence on cord or brain, are those of the upper parts of the limbs—those connecting the legs and the trunk—there is not the irregular movement of the feet which you are accustomed to associate with pronounced tabes, not the high step and sudden descent of the foot, but there is the unsteadiness which you are accustomed to associate with disease of the cerebellum. It is a defect rather of the maintenance of equilibrium than of the movements of the legs as a whole.

I now test the knee-jerk. You see that it is absent. This is proof that there is an interruption of the path from the muscles to the spinal cord—that the common mechanism of ataxy is here present. The man is a plumber by occupation, and there is in his past history one other very important fact, namely, that syphilis can be excluded. Unless we could have excluded syphilis we should not have been justified in attributing this to lead, even though we have some of the symptoms of lead poisoning. The mere exposure to the risk of the contraction of syphilis is found in so many cases to be the only antecedent of unquestionable syphilitic symptoms, that we cannot exclude syphilis unless we can exclude the common mode of its contraction. The hypothetical water-closet is, in my experience, purely hypothetical, and need not be taken into practical consideration. The only exception to the common mode of infection is presented by those who attend labours. In this case, then, we can exclude syphilis. He has been a plumber since the age of fifteen. There is no other cause which we can trace; although he has never had wrist-drop or lead colic, and yet he has a distinct lead line on the gums. These two facts are important. You know wrist-drop and colic are common effects of lead poisoning. But here we have evidence of the presence of lead in the system in a man who has worked in lead for twenty-one years, and yet has not had the common symptoms. From this we may infer that, either owing to personal cleanliness or for some other reason, the lead has passed into his system in small quantities or in unusual combination, and these are the conditions which I have indicated as those to which we must look for an explanation of the occasional anomalous effects of these poisons. So that, you see, the very fact that he has been exposed to the influence of lead, and yet has never had the common consequences, makes it easier for us to believe that this uncommon consequence is the result of the lead poisoning. He has, moreover, no affection of sensibility, but he has had pains not unlike those of ordinary tabes. He has no obvious loss of power in his legs, and yet he has a symptom which at first sight seems to indicate loss of power. He has, besides the defective power of maintaining equilibrium, a very peculiar difficulty in rising from the floor. He can only do so when he helps himself by putting his hands on some adjacent object. But he does this, not to aid the muscles, but to guide the movement. It is the result of the localisation of the inco-ordination to the lower trunk muscles, and to the muscles which connect the trunk and the legs. In his case it is the symptom which simulates weakness; when he is lying, all those muscles are able to exert full power. It is of great importance to be aware that simple but considerable inco-ordination in those muscles may make the assumption of the erect posture difficult. When he rises a second time, knowing the reason for his difficulty, you will note that, from the very suddenness by which he manages to get up, there is no defect of power. He does it now much better than he did when he was admitted to the Hospital. A few months ago he could not get up by himself, and now, sometimes, he can just manage to do so.

One other point that you should note is this—there is no defective action of the pupils such as you so often find in true tabes. Although too much weight must not be attached to it, the absence of this symptom is always some reason for thinking that tabetic symptoms are due to some other poison than that which usually causes them.

He has not much lead line, and I should like you to notice that it is only to be seen where the gums are detached from the teeth. I have before explained to some of you the way in which the lead line is produced. When the gum is in perfect apposition to the teeth there is no lead line. The lead line is sulphide of lead deposited just beneath the surface, outside the tiny loops of vessels—a disposition which gives it a granular aspect under a lens. A layer of such sulphide of lead is deposited beneath the inner surface of the detached gum, and it is the edge of this layer that is seen as a line. The sulphur comes from decomposing albuminous materials lying between the tooth and the detached gum. Unless albuminous materials lodge there, no sulphide of lead is formed, and there is no lead line. I have seen cases of extreme lead poisoning, with perfect gums, and with no trace of lead line. But yet, as most people have somewhere a space between the edge of the gum and the teeth, sulphur-yielding materials accumulate there and form at least a fragment of lead line. It is instructive here to notice that the line is confined to those two teeth in which the gum is detached, and that against the others, to which the gum is in close apposition, there is no line.

The treatment of such a case is, first, the ordinary treatment for lead poisoning; and, secondly, when the toxic influence has been entirely removed, we should attempt to help the processes of regeneration by agents, such as strychnine, which seem to promote the nutrition of the nerve elements. I will, however, only add this injunction : Remember that all these metallic poisons become stored up in the tissues as compounds with organic substances, and the first effect of releasing them from the tissues is to increase the quantity in the blood. Give iodide of potassium in lead poisoning and the amount of lead is increased in the urine. Iodide of lead is insoluble, but the lead forms a soluble combination with the iodide and albuminous bodies which can pass from the tissues into the blood and away by the excretory organs. At first the quantity of lead in the blood may be very much increased, and its resulting effects also increased. Some of you will remember a remarkable case of arsenical poisoning with nervous symptoms and rash, which I showed last summer. Iodide has a like effect in arsenical poisoning. In that patient the iodide had to be discontinued, so greatly were the symptoms aggravated by the irritation of the arsenic which the iodide brought from the tissues into the blood. These symptoms continued for a long time, but now the patient is practically well, except for some pigmentation which still persists. The recovery from these effects is slow; happily, however, it is, as a rule, sure. As the lead passes from the nerve elements their vital power of nutritional renewal entails a slow restoration of structure and of function, and although, in this patient at present, improvement is slow, it is definite, and we may anticipate that it will progress to restoration of the normal state.

LECTURE XIX.

OPTIC NEURITIS.

I.—SLIGHT.

Gentlemen:—My task to-day is to endeavour to help you to help yourselves. There are some things that may be taught by another, but there are many things which the learner can only teach himself, and it is with one of these that I want to occupy your attention to-day. I will do so for less than the usual space of time in order that you may have time to learn by observation that which you can only learn in this manner. And I do so because I happen to have the opportunity of affording you the means of observing a case which is almost unique in its characteristics and in the instructive points that it presents. There is one symptom only to which you need attend. The other symptoms are so uncertain in their significance that it is doubtful whether it is worth while for me even to bring them before you for their negative significance.

The symptom which is presented is optic neuritis. The unusual features which it presents are that it is neuritis in one eye only, and that in this eye the neuritis is partial. Any of you can come at another time and examine the patient carefully at your leisure.

You probably know that it is rare for optic neuritis to be one-sided. It is one of the symptoms which usually occur on both sides at the same time, because, as a rule, its cause is either an influence from the brain, which acts

---

Delivered October 24, 1894. *Clinical Journal,* December 5, 1894.

equally upon the two sides, or an influence from the blood, which likewise has an influence so general that its effects are bilateral. Whenever you can compare one side that is normal with the other that is diseased, whatever kind of disease it may be, you are always at much greater advantage in your observation than when there is perfect bilateral symmetry in the morbid change. Many slight morbid changes are difficult to recognise in their absolute degree, and yet not difficult to recognise if there is the opportunity for comparing an absolutely normal state on one side with the slight abnormal condition on the other.

The case is instructive not only on account of these features, but because the neuritis has developed in a preexisting condition that is itself abnormal—a condition which depends upon an extreme degree of myopia. This fact adds to its practical utility to advanced students, because a considerable degree of myopia increases the difficulty of observation and the risk of error; and thus the careful study of the case will not only give you definite information, but will also increase your ability to overcome a difficulty which is not uncommonly met with in ophthalmoscopic work, observation under the condition of considerable myopia. As you probably know, this difficulty only obtains in the direct method of examination, in which there must be accurate compensation by a lens behind the ophthalmoscope. I fear I seem to be making the advantages of the case depend on its difficulties, but, in point of fact, every great difficulty that is overcome lessens a host of minor difficulties. I may further press on you the necessity for careful estimation of the conditions under which you are observing, inasmuch as the myopia here is almost twice as great on one side as on the other. If you try to compensate on one side in the same degree in which you compensate on the other, you will be unsuccessful. I therefore impress upon you the necessity, in the first place, of always observing and ascertaining what are the conditions under which you are making your observation.

This necessity is very important in all diseases. Students, and I fear sometimes those who are no longer nominal students, listen to the apex of the heart where the apex of the heart ought to be, instead of at first ascertaining where it is. I have frequently known a student fail to find a murmur at the apex of the heart, because he was listening where the apex ought to have been, not where it was. The patient who is before you has a cardiac condition worthy of observation, although it will not involve that difficulty. There is a loud murmur at the apex, but it extends so widely that you will not miss it, even though you miss the apex and have not done what you always should do—feel where the apex is before you listen. It is a systolic murmur, and there is also a systolic murmur at the base, which is certainly a separate murmur. Never omit carefully to observe collateral symptoms. You will examine this case on account of the condition of the eyes, but it can teach you a good lesson in auscultation of the heart; and if you have another opportunity of examining the patient you should make a careful examination of the heart. When there is a systolic murmur at the apex and a systolic murmur at the base, the one due to aortic stenosis and the other to mitral regurgitation, it is generally not difficult to ascertain the fact that there are two murmurs if you are sufficiently careful to observe the gradation of the sound, or, rather, the sequence of reversed gradation, in passing from the apex to the base. The pure mitral murmur becomes fainter; if it does not, and is even louder at the base and over the aortic orifice than at the apex, you may feel sure that there is aortic constriction, and especially if, as in this case, you can observe a change in the quality of the murmur. I mention this because the co-existence of mitral regurgitation and aortic stenosis, without aortic regurgitation, is not very frequent, and whenever it is met with it should be most carefully observed, because it is not always, for the reasons I have mentioned, easy to detect.

I said that the difficulty involved in myopia is chiefly in the direct method of examination, but a fallacy is caused by myopia in the indirect examination. In this method, the optic disc looks small in myopia and large in hypermetropia. Just as you should always, before listening to the apex of the heart, find out where the apex is, so, before looking into the eye, you should ascertain roughly what is the state of refraction. That is easy to do. First throw the light into the eye by the mirror from a distance, from the distance you would adopt in the indirect method of examination. Move your head a little, and you will quickly catch sight of a vessel. If you see that vessel distinctly, the refraction is not normal; there is either myopia or hypermetropia. Move your head again a little to one side or the other; if the vessel seems to move in the same direction, there is hypermetropia; if it moves in the opposite direction, there is myopia. You then know what to do when you proceed to the direct method of examination, and you know how to estimate the size of the optic disc as it appears to the indirect method of examination.

This case illustrates the fact in a curious way. I show you large, somewhat rough drawings of the discs, and I have here also a more careful drawing which you can compare with what you see when you examine the patient (figure). In the left eye (on the right, in the figure) the optic disc, of oval outline, appears larger than you would expect. It looks larger than the normal disc, although the eye is myopic, and in myopia objects should look small. It is not the optic disc you see. Note that there are two oval outlines; the optic disc is the inner area, and the outer boundary is simply the limit of what is called a "posterior staphyloma;" that is, an area of atrophied choroid, so common in myopic eyes in consequence of the greater length of the eyeball. The tendency to increase in length leads to a condition in which the opening in the choroid does not reach the opening in the sclerotic, and is increased by a process of atrophy of the choroid.

This case has still further the utility that the arrangement of the vessels is particularly puzzling, and if, with the help of the drawing, you discern them perfectly, you will have enabled yourselves to overcome many other difficulties which you will meet with afterwards in other cases of peculiar vascular arrangement. The peculiarity here is that the central artery is so situated as to entirely conceal the origin of the veins. In the left, the normal disc, you see in the middle the white area of the physiological cup, and the vascular portion of the disc extends up to the place at which the vessels emerge. If you are familiar with normal eyes you will know that that is very frequently the case, especially when the physiological cup is large. The dark portion on the other side of the cup is greyish in tint and not vascular. I have purposely had these diagrams done in black and white and not in colours, for the same reason that I had most of the drawings in my "Medical Ophthalmoscopy" done in monochrome, because it is of considerable more importance to attend to the changes of form than to the change in colour. Nine-tenths of that which is important in the disc depends on change in form—on that which can be shown in black and white, and not in change in colour—and nine-tenths of the errors that are made in interpreting the appearances are owing to undue weight being given to changes in colour. I should hardly like to say how many cases of normal discs I have had shown to me, as the seat of optic neuritis, simply because the discs were highly coloured. The vascularity of the disc varies as much as does that of the cheek. The test of a morbid condition is not the colour, it is the change in form,—it is especially obscuration of the edge.

In this optic disc, the arteries, as I say, happen to be so placed that they entirely obscure the portion of the veins where the branches unite to pass into the nerve, and it is only on very careful observation that you can discern which

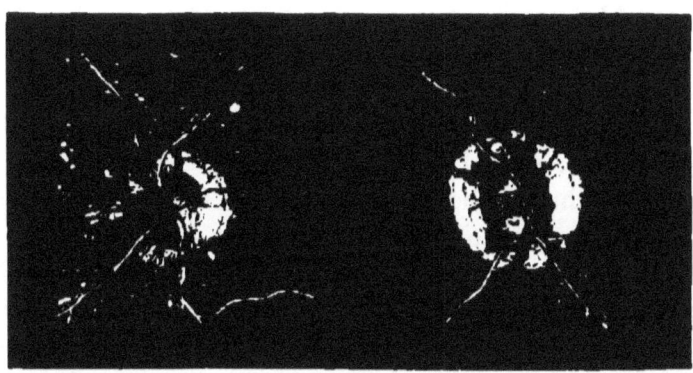

Partial optic neuritis in the right eye only. Both eyes are myopic. In the right (the patient's left) there is a posterior staphyloma extending all round the disc. The cup is steep on the nasal side (to the right in the drawing), which is of the inverted image, so that in this disc the side further from the other eye is really the nasal, and the darker tint on the sloping side of the cup (apparently towards the other eye, but the temporal) is due to pigment. The emerging mass of nerve fibres extends up to the vessels on the other side.

In the left eye a similar condition is seen in rather less than half the disc; that apparently towards the other eye and therefore really away from it, *i. e.*, the temporal side. The side further from the other eye, that is, the real nasal side, is obscured by inflammatory swelling, which presents also some small hemorrhages. The clear side, really the temporal, is of course the side turned towards the yellow spot, in which the nerve fibres are few or absent. The figure shows the tendency of neuritis to develop first where the nerve fibres pass off and the interstitial tissue is abundant.

are arteries and which are veins. The veins seem to arise from the arteries. It is always necessary to look carefully, to be quite sure, especially in the case of smaller vessels, which is an artery and which is a vein. It is the colour and the character of the central reflection that distinguish them.

Turning to the other eye, we see a condition which is also most instructive. Here also the venous branch is obscured at the beginning by an artery. The upper vein is the most conspicuous object that attracts attention. On looking at the disc, you will be struck by the fact that at one side there is a pale, or almost white zone, pretty sharply margined on the outside, and limited inside by a redder zone. The outer white zone is a condition of choroidal atrophy, revealing the sclerotic,—a " posterior staphyloma," similar to that which is seen in the other eye. The disc itself begins at the inner oval. As the figure shows, the edge of the disc is distinct on one side, but is entirely lost upon the other. There, you can see no margin to the disc, and you can see no zone in which the white sclerotic is visible in consequence of atrophy of the choroid. It is all covered by a faintly striated, red tissue. On the inner part of this there are several small but distinct hemorrhages. The amount of swelling is slight. Swelling can be estimated accurately by ascertaining what strength of convex lens behind the ophthalmoscope is required to make indistinct the surface of the swelling compared with the adjacent retina, or, in the case of myopia, what difference in the strength of concave lens is required behind for distinct vision of each. That is, however, a difficulty in myopic eyes, and you can generally, without much difficulty, estimate the amount of swelling in two other ways. First, whenever there is much swelling, the central reflection of a vein is lost where it descends the side of the swelling. You know the cause of the central reflexion from the middle of a vein. That which we call a retinal vein is not the vessel, but the column of blood within it; and from the centre of the convex column of blood the light is reflected straight back to the eye, whereas from the lateral parts the light is reflected at an angle. The consequence is that all the light is reflected from the central part of the convexity, and this prevents the colour of the blood being perceived as at the sides.

But supposing a vein slopes for a short distance, so that its course is in a plane at an angle with the direction of the light other than a right angle. The central reflection is only seen where the vein is running in a plane perpendicular to the line of sight. In a part at which the vein slopes, there is no central reflection, and so the vein appears darker, because in the centre the colour of the blood is seen as it is at the sides. Therefore, whenever there is the appearance of swelling of the disc, observe most carefully the veins as they pass on the sides of the affected area, and notice whether there is loss of the central reflection. If there is considerable swelling there is sure to be this loss. In this case it is scarcely perceptible, and therefore the amount of swelling is slight. The other method by which the amount of swelling can be easily estimated ought to be familiar to you, because it is of such wide application to every condition in which there is a difference in the level of objects in the fundus; as, for instance, to the depth and shape of the "cup." It is that which is called "parallactic," because, by an almost presumptuous use of a word which has a wide and important meaning to designate an exceedingly small subject, the word which describes the means of ascertaining the millions of miles of distance of the earth from other planets is applied to ascertain whether one object in the fundus of the eye is a millimetre in front of another. It depends simply upon the fact that the relative position of two objects is different if you change your point of view. In the case of the ophthalmoscope,—if, in the direct method, you move your head a little to one side and back again; if in the indirect method, you move your lens in the same way,—you see a change in the relative position of the objects. In astronomy it is necessary to wait until the other side of the earth's orbit is reached to observe the change in the relative position, but the principle is essentially the same. Never omit an opportunity of practising it in ophthalmoscopic work. Whenever you look at even a normal disc, move the lens a little, or move your head in the direct method, and you will be surprised how the relative change of position of objects at the edge of the disc and at the bottom demonstrates to you their relative distance almost as perfectly as if you were looking through a stereoscope.

The great fact of optic neuritis is the blurring of the edge of the disc in consequence of increase of tissue in front of it. The opacity of the tissue prevents the normal reflection of that which is behind, which can pass without difficulty through the quite translucent normal structure. Remember that the blurring always begins upon one side. You know that the nerve fibres which come from the optic disc, and radiate on the retina, come off on the nasal side, and above and below, but scarcely or not at all on the temporal side, towards the yellow spot, in spite of the fact that the yellow spot is the centre of vision. Hence it is that, at that part of the disc, the structures are absent, the nerve fibres, and the tissue between them that is the chief seat of inflammation, present the first visible signs of neuritis. When it is intense and considerable, there is sufficient tissue, even where it is least, to permit inflammation to invade the whole periphery of the papilla; but when it is slight the changes are limited to the side on which they are first seen. This is well seen in the figure before you and in the patient from whom it has been drawn. I am anxious that you should carefully observe it. The inflammatory change obscures the edge above and below, and on the whole nasal side, while it is absent on the temporal side.

In normal states there is always such transparency of the structures in front of the edge of the disc as to enable you to see it, even where the nerve fibres are most abundant. The first effect of neuritis is to increase the obscuration by making the tissue which is over the edge less translucent, and so to cause the apparent "blurring." Thus indistinctness of the edge is the first manifestation of the process. But the edge may be indistinct from another condition. You may have much difficulty in seeing it because the outer part of the disc is so red as to be of the same tint as the choroid. It is through this that the mistake is often made to which I have referred, the mistake of thinking that a disc that is simply and naturally red, is inflamed. This important distinction, however, is not difficult. The rule that you should fix in your minds is this. Let me first explain its reason. When you look with the indirect method, the degree of magnification is low, the illumination wide, and focus deep; when you look with the direct method, there is a much higher degree of magnification, and a much more limited illumination and shallower focus. It is so with the microscope; a low power affords a considerable depth of focus; with a high power you have a very shallow range of focus. Hence, when you use the ophthalmoscope by the direct method, with its high magnification, you have a very shallow layer of focus compared with that which is afforded by the lower power of the indirect method. Hence it follows that if there is anything in front of the edge of the disc, obscuring it, when you examine by the direct method this layer is focussed, and not that which is behind it; the latter is not perceived, partly as it is not in focus, partly as that which is before it is in focus and so is distinct. With the indirect method, however, a lower magnification, a deeper focus, and a greater illumination enable us to see distinctly the tissue in front of the edge of the disc and also that which is behind this tissue, the edge of the disc itself. With the direct method, real obscuration is greater than it is with the indirect. But if the edge is indistinct from resemblance in tint of the disc to the adjacent choroid, when the eye is examined with the direct method the edge of the disc is *more* distinctly seen. Thus there follows this most important rule, that, if you think there is commencing neuritis, when the appearances are more natural with the direct method than with the indirect, it is normal; they are more morbid when there is more obscuration with the direct method. This is a rule of great practical importance; you may trust it, and if you endeavour to apply it you will soon discover its value.

I intended to describe to you the symptoms which the patient presents, but I think it is not worth while to do so, because the case is one in which a precise diagnosis is impossible, and in which the process of attempting to make a diagnosis is not of such conspicuous use as to justify me in diverting your thoughts and over-crowding to-day's memory. She has slight weakness on one side, a history of some giddiness, and of some headache of varying degree. A little weakness is thought to exist in the external recti, but it is not conspicuous. There is no nystagmus, and if it were not for the optic neuritis I think we should scarcely be inclined to consider that there was any organic disease. It is one of those cases in which there may be a small slowly growing tumour in some situation, or a tumour which is stationary, or increasing only at intervals as a tubercular tumour may do. But the symptoms themselves do not indicate tumour, nor does the optic neuritis.

The last statement may surprise you. The chief and first significance of optic neuritis is the presence of organic disease. If you can exclude a blood-state, or constitutional condition, you may feel sure that there is organic change within the skull or orbit, such a change as could be recognised almost always with the naked eye, and always at any rate with the microscope. The organic brain disease causing neuritis is of an irritative character. Irritation is the one element with which we can connect the process of neuritis, but the questions of mechanism and causation I must deal with upon another occasion. The point I wish now especially to press upon you is that, when neuritis is due to intracranial organic disease there is a relation between the course of the morbid process in the brain and the visible inflammation within the eye. Hence, neuritis is important, not only for diagnosis, but for prognosis. Many times I have known the commencing subsidence of neuritis to be the first indication of the commencing subsidence of a morbid process in the brain, and I have known the persistence of neuritis to be the indication of persistence of disease in the brain, of which every other effect, for the time being, had passed away. In each instance, and each class of instances, the indications have been absolutely verified. Further, chronicity indicates chronicity: chronicity of the neuritis, chronicity of the process. The converse is not always true. A chronic process in the brain, going on slowly, sometimes causes an acute development of neuritis. Doubtless the exception is only apparent; some acute consequence of a chronic growth causes the acute neuritis. A rapidly growing tumour never causes a chronic neuritis.

Neuritis is, however, sometimes so chronic that you may watch for month after month, and observe no change, and then, if there is organic disease, you can be certain that it is of a most chronic character. That is the case with this patient. I say, "If you can be sure of organic disease," because I have seen cases of slight optic neuritis of extreme chronicity in which, although there was no constitutional condition to explain it, I could not be sure of organic disease. Indeed, I have sometimes felt sure there was no organic disease. Those cases at present are a mystery. In a few cases, during a year or two years, I could see no change in the aspect of the discs, and in one of them I came to the conclusion that it was a congenital condition. But conclusions, gentlemen, seldom merit their name. Most conclusions should be regarded as beginnings, and not as endings, if you can pardon the paradox; it happened that five years afterwards I saw this patient, and every morbid appearance had passed away. It is possible that hypermetropia aids in the development of the condition. But all the patients, to whom I refer, presented evidence of functional disturbance of the brain, and I think there are cases in which very chronic optic neuritis is due to the combined influence of what we call "functional" brain affection and hypermetropia. How these causes act is still a very open question. But in none of those cases was the process sufficiently active to lead to extravasation of blood, and although in this patient, during the four months she has been under observation, no change has been perceptible in the optic neuritis, there have occurred the small extravasations of blood, which show an activity that is definite and seems to take the case out of the category of those that I have just been referring to. (Here again I have to guard the statement.) I say "seems," because the features suggest the question whether the occurrence of the extravasations may not be specially related to the peculiar arrangement of the tissues and vessels, and therefore whether it has so much meaning as we can generally ascribe to them. Remember that exceptional facts are often related, and that an unusual condition may be the cause of an unusual feature—that the preceding, congenital state here may be the reason why there are such hemorrhages as are scarcely ever observed in neuritis that is at once so slight in degree and so extremely chronic in its course.

It will be more useful for you to carry away, instead of a confused perception of the possible meaning of the other symptoms presented by the patient, a clear perception of this one fact. There are cases of optic neuritis in which the process has not its common significance. There are cases in which it does not even help us to say that organic disease is present, in which it leaves the diagnostic problem very much as it would be without the neuritis. It should, indeed, invariably have one effect upon our minds. Like many other symptoms, in many other morbid states of various kind and in every part, it should make us more watchful, more careful to maintain our observation, and more ready to give due weight to any unequivocal symptom that occurs,—remembering ever that symptoms that are unequivocal alone may be decisive when combined.

LECTURE XX.
OPTIC NEURITIS.
II.—Severe.

Gentlemen :—A few weeks ago I drew your attention to a case of optic neuritis of peculiar instructiveness on account of its slight degree and unilateral character. One eye was normal, the other presented the least degree of definite neuritis on one half of the disc. Although the nature of its cause was uncertain, we could not doubt that the changes were due to an intracranial process. I may add that the condition has since undergone little change, sufficient to show no more than the fact that it is not stationary.

To-day I wish to take an opportunity, although it is an imperfect opportunity, of presenting to you the features of a case of optic neuritis which stands at the other end of the scale. This, also, is due, without doubt, to intra-cranial disease; but instead of being slight in degree, it is one of the most severe I have ever seen; indeed, it is, on the whole, the most severe I have seen. I am, therefore, unwilling to lose the opportunity of impressing on you the features of this important symptom in its extreme degree. Although the patient is before you, I am sorry that I cannot invite you to examine her as you examined the other patient. It may be practicable for you to come at some other time and look at her eyes, but she is enduring such severe paroxysms of pain that even movement down into this theatre is as much as it is right to subject her to. But I

Delivered January 16, 1895. *Clinical Journal,* June 5, 1895.

will tell you the salient facts of the case, and the mere sight of the patient will impress them upon you.

She is a young woman, aged twenty-two, who has lost a brother from consumption. Last June, seven months ago, she began to suffer from headache. It was at first slight, but quickly became severe; partly frontal, and partly occipital. Transient weakness of the left internal rectus was observed. The headache and the diplopia led to an ophthalmoscopic examination, and the observer, a country practitioner, found, to his surprise, intense optic neuritis; so intense and considerable that it must have either developed with extreme rapidity or must have existed for some time before the headache began. The neuritis suggested a tumour, and she was treated with large doses of iodide of potassium. The common procedure, whenever there is reason to suspect an intra-cranial growth, is the administration of iodide of potassium, because a growth is sometimes due to syphilis. This treatment is adopted rightly when this cause is possible, but occasionally without adequate perception of the fact that it is impossible, although this fact, sometimes, be clearly discerned. The treatment increased her discomfort. I cannot say that it did her definite harm, yet you should know that this treatment may have, indirectly, disastrous results. The most intense optic neuritis, precisely like that of cerebral tumour, may be due to anæmia, at any rate to the chlorotic anæmia of girls. I have seen a case of such neuritis which was only discovered when sight had become gravely impaired, and then, although syphilis could be absolutely excluded, the case was treated with iodide of potassium until all sight had vanished. In the case referred to, I am certain that the prompt administration of iron, combined with perfect rest, would have secured the recovery of a fair degree of sight if not of perfect vision. I have known cases in which, I am sure, sight has thus been saved. I mention rest, because, in what may be called a critical state of anæmia, it is of the utmost importance that the deficient hæmoglobin should be saved for vital processes. In these cases, so different from pernicious anæmia, the defect is in the hæmoglobin, in the element that carries oxygen to the tissues. The first thing is to prevent the inadequate supply of oxygen being wasted in the release of energy from the muscles. Without adequate oxygen the machinery of life cannot be maintained in the needed power. The first essential is to save for this the oxygen it would waste in energy. That patient has now been for several years, and will be, as long as she lives, absolutely blind.

Thus, you will perceive, that, while the administration of iodide of potassium cannot have, in such cases, much beneficial influence, and, probably, has none, it is not unimportant because it hinders the adoption of proper treatment. It is doing something, but this therapeutic refuge does not always illustrate the maxim that " something is better than nothing." It may be no worse than nothing, yet there may be another comparison—there may be something else which might be done, and which would save that which, to many, is more precious than life itself. If it is, in general, true that " knowledge comes and wisdom lingers," we must, alas, admit that knowledge is too slow of foot and is too long in spreading. That which is not often needed, not often enforced by observation, however simple it may be, however practical, seems to find no holding-place, and when urgent need arrives, it is not there.

But this is " by the way." We must leave it and come back to our patient's history. Her medical history is somewhat of a blank for a time, save for the fact that her sight slowly failed. A month or six weeks ago she could just count fingers with one eye, but since then she has lost all perception of light.

She now comes here with headache, paroxysmal, but

most intense. Up to about three weeks ago the headache had been but slight, but during the last three weeks it has become much more intense, although still paroxysmal, and it has been more prominently occipital, and more on the left side than on the right. During the paroxysms there has been a little retraction of the head. Once or twice vomiting has been associated with it, but vomiting has not been a conspicuous symptom. She is able to stand, but shows a little unsteadiness. There is a history of occasional slight attacks of subjective sensation of " numbness " in the left arm. There is no foot-clonus; the knee-jerk is excessive. There is not now any distinct difficulty in the movement of the eyes, or paralysis of any cranial nerves.

With such a headache, combined with optic neuritis, it is impossible to doubt that she is the subject of an intracranial growth. These indications are the more definite because the headache is occipital. Optic neuritis in anæmic girls may be associated with very severe headache, but this is general or is greatest in the front of the head. It is also more or less constant—not paroxysmal, as in our patient. Moreover, what is of great practical importance, there is no sign of definite anæmia, which, were this optic neuritis of anæmic origin, would unquestionably be conspicuous.

You can see, even from a distance, the blind look of her eyes. You can see the large pupils which, I may say, do not act to light, and you can also see the slight retraction of the head. If, as we know, there is optic neuritis of great degree, the absence of reaction of the pupils becomes unimportant. But if a patient has optic neuritis in very slight degree, insufficient to impair sight, the absence of any pupillary action would show conclusively more considerable morbid process in the optic nerves, or at the chiasma. Here the changes within the eye are ample to account for the loss of sight and of action of the pupil.

It is to these changes that I desire to draw your attention. Although you cannot to-day observe them for yourselves, I am fortunate in being able to put before you two admirable drawings of the fundus, by Dr. Ridley (reproduced in the accompanying figures). In the left eye you see a pale area, not much bigger than the disc, soft-edged, and distinctly prominent, as the curves of the vessels show, and as is also shown by the change in their central reflection. But there are also changes in the retina, white spots and flakes and streaks, not only around the macula lutea, but on the opposite side of the optic disc. Beyond the side of the papilla, which is away from the yellow spot, there is a conspicuous series of such flakes, unusual as a consequence of neuritis. Note also that between that and the disc there is a curved line, lighter than adjacent fundus, concentric with the disc. Such a line is seen occasionally in cases in which the optic neuritis has been very much more extensive than at the time when it was observed. It is due to the swelling of the substance of the papilla, which, pushing away the adjacent retina, has thrown it into a fold. You will find illustrations of the process in "Medical Ophthalmoscopy." I have seen two, or even three, such folds under the microscope. During the height of the inflammation they are inconspicuous, but they remain when the process has subsided. We shall presently have to note their earlier stage. In consequence of the reflection of light from the convexity such folds appear as pale lines.

This condition indicates a neuritis which has to a considerable extent subsided. At a previous period the papilla no doubt presented very much the aspect which the other papilla now does. Although neuritis is usually double, it frequently develops somewhat earlier in one eye, and often reaches its greatest intensity in one eye sooner than in the other. So it is in this patient. The one eye is in the stage of subsidence, while in the other the process is at its height. To this, as shown in the other figure, we will now turn.

PRIMARY PAPILLITIS.

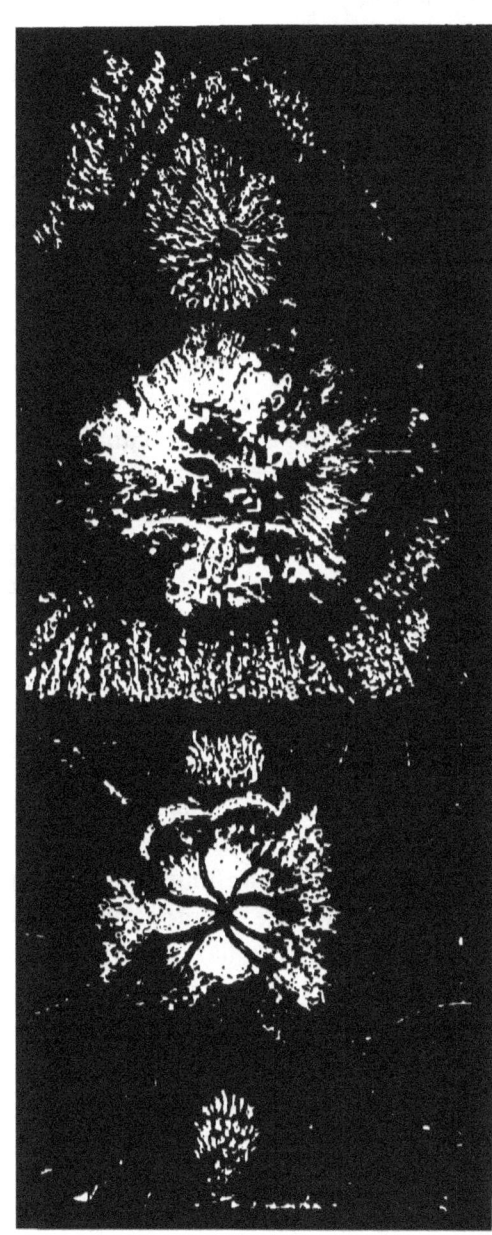

SEVERE OPTIC NEURITIS, WITH SECONDARY RETINAL CHANGES. (*From Drawings by Dr. Ridley.*)

Fig. 1. Left Eye. A soft-edged papillitic swelling (too pale and sharp in the figure), the prominence of which is shown by the curves of the vessels, and their reflections. Descending the slope of the swelling, they are then partly obscured by the swollen retina into which they pass.

Fig. 2. Right Eye. Wide papillitic swelling, with extensive changes on the retina (see text). This in the figure is not pale enough in tint; it was chiefly due to œdema (see *Postscript*).

The appearance in the right eye in which the condition is in the acute stage is such as I have never seen before from intra-cranial disease. Indeed, I have never seen the like from any cause; I have never seen such extensive damage to the retina in any other case of primary papillitis, such as this undoubtedly is. In the position of the papilla is a wide prominence four times* the width of the normal optic disc. (The photographic reproduction has rendered the swelling too dark.) You will notice that its soft edge is broken by radiations, at which the whitish tissue extends further on the retina. It amounts in the left eye to four diopters, and probably to six in the right. But the most remarkable feature is the extent of damage to the adjacent retina. I called your attention in the left eye to the white dots and flakes about the macula and on the other side of the papilla. The same are seen in the right eye, but in this they are arranged all around the macula lutea in a double range, and also around most of the circumference of the disc. On the side opposite to the macula they extended beyond the area that could be brought into the figure.†

In " Medical Ophthalmoscopy " you will find a figure of a case of intense optic neuritis which had involved the retina to as wide an extent in the active process, as it is here involved in the inflammatory swelling, but the drawing was made after the subsidence of the inflammation. Around the macula, white spots and flakes were left, very nearly the same as those shown in the drawing of the right

---

* The gradation of the edge of the right neuritic swelling is too sharp and the swelling is too white in the figure, which is, indeed, a very inadequate reproduction of the excellent drawing of Dr. Ridley. The actual diameter of the disc would not be more than three-quarters of that of the pale area.

† An important paper on the relation of these flakes to the structure of the retina and to œdema, by my colleague, Mr. Marcus Gunn, will be found in the " Transactions of the Eighth Ophthalmological Congress," 1884, p. 77.

eye of this patient. The figure in "Medical Ophthalmoscopy" is from a case of optic neuritis due to anæmia, not to an intra-cranial cause; it was primarily an inflammation of the papilla and not of the retina, but involving the retina secondarily.

The fact that these changes in the retina resemble those of albuminuric retinitis is one of very great practical importance. A case such as this, in which each eye presents a subsiding neuritis, with changes about the macula lutea of this character, may accompany symptoms of intra-cranial tumour which are not decisive, and there may be also some albumen in the urine. You can imagine how difficult it would be in such a case to decide whether the changes were due to kidney disease or were the result of previous intense neuritis from an intra-cranial cause. The difficulty is not hypothetical. I have met with the combination and with this difficulty, especially in young women and in children. The albuminuria has been slight; there has been no evidence in the urine or elsewhere, of structural kidney disease, and there have been symptoms of intra-cranial tumour, suggestive though not conclusive. But in the cases in which the retinal changes are secondary the papilla always presents the features which indicate a considerable previous inflammation. Knowing that such neuritis will lead to these changes you should have no difficulty. It is only when the neuritis is slight, and there is no great amount of new tissue in the papilla that albuminuric retinitis should be thought of.

But there is another side to this question. Albuminuria may cause neuritis and not retinitis. It may cause limited optic neuritis, in which there may be not only changes in the retina that are so slight that you have to look for them most closely by the direct method, but in which there may be no changes whatever in the retina—a simple optic neuritis, never anything like so intense as that which is seen here, but which may still be considerable in degree. A further inconvenient fact for us, and probably a significant one, is that when an isolated optic neuritis is a consequence of renal disease, there is generally headache as a very pronounced symptom, so that the symptoms may be very equivocal in character. I remember having this fact impressed upon me at the beginning of my work with the ophthalmoscope in medicine.

The case was seen when I was a student at University College Hospital, and first yielded to the fascination of the ophthalmoscope, which was then in its early days as regards its application to medicine. The patient suffered from severe headache and repeated convulsions. There was moderate but distinct optic neuritis, with no change whatever in the retina. I ventured to make a confident diagnosis of cerebral disease, but the post-mortem examination showed only granular kidneys.

The last point of practical importance suggested by the case we have considered is this: In the remarkable series of surgical operations, which have been done here by Mr. Horsley, it has been found that, in cases of intra-cranial tumour, and in which nothing could be done to remove the tumour, trephining lessened the headache in a remarkable degree, and led to rapid subsidence of the optic neuritis. Special attention has been drawn to this fact by my colleague, Dr. James Taylor. This can only be explained on the assumption that a causal influence is removed by the operation in the diminution of the intra-cranial pressure. The significance of the fact as regards the mechanism by which optic neuritis is produced may be exaggerated. We know that great excess of intra-cranial pressure may exist without optic neuritis. It is probable that its influence is to intensify and prolong a neuritis that would otherwise be slight and brief.*

---

* See, however, added note.

But this raises the special question—which is, too, a somewhat new one—whether an operation in this case would be justified in the hope that it might induce a subsidence of the neuritis and save some sight, apart from any attempt to remove the tumour. For the latter purpose the question is beset with great difficulties.

It is one of the cases in which the diagnosis can only be carried to a certain point. In the first place, the pain, although it is to a small extent frontal, is chiefly occipital, and has a tendency to pass down the neck. This is generally an indication of subtentorial disease. But there is also some tenderness above the level of the tentorium on percussion of the skull. Percussion of the skull is of some use in localising disease. I do not, however, think that we have, at present, sufficient grounds to say whether the trouble is in the occipital lobe, or whether it may not be disease of the cerebellar hemisphere. The tenderness is so conspicuously greater on the left side than the right that it suggests a unilateral seat of the disease. If it were, in the occipital lobe, we should expect to have had a history of hemiopia before sight failed. Yet the history does not permit us to assume that this was absent because its record is absent. In the next place, is there any reason to get at the case, excluding glioma and tubercle? If there is a family history of phthisis; if the patient is in the first half of life; if the cause is neither very acute nor very chronic, tubercle is most probable.

In conclusion, I desire to say one word on the practical question. Is there any chance of benefit from the operation of trephining?

The significance of the fact that trephining allows the optic neuritis to subside may be easily overrated. It seems to show that the increase of intra-cranial pressure is the cause.

To put it shortly, I believe that intra-cranial pressure is a factor which intensifies the neuritis; you see its evidence in the swollen sheath behind the eye in most of these cases. But there may be optic neuritis without a swollen sheath,* and there may be intra-cranial pressure without optic neuritis. I remember a case, which shows to how great an extent intra-cranial pressure may exist without optic neuritis. Over the right hemisphere there was growing from the dura mater a sarcoma of the size of my fist. The brain gave every evidence of compression. The patient had been in the hospital, where we had watched her. I was then Registrar, and I watched the case afterwards in one of the suburbs. The case is fixed in my mind, because I did something which afterwards alarmed me. The sun was shining in at the window towards which the head of her bed was placed. I used the sunlight for the ophthalmoscopic examination, and I was afraid afterwards that the concentration of such a powerful light might have blistered the retina. But the patient died too soon to permit any effect to have been important.

Intra-cranial pressure, however, may prevent the escape of the products of inflammation from the papilla, and perhaps may cause additional morbid material to reach the papilla from the meninges. It may thus intensify the optic neuritis. I will not say that it may not possibly excite it, but I believe the chief element of excitation is the passage down the nerve of slight interstitial inflammatory changes. It is possible that, if there were no increased intra-cranial pressure, in a great many cases the optic neuritis thus produced would be slight, and perhaps the cases in which trephining, by relieving the intra-cranial pressure, allows the neuritis to subside, are those in which, without the increased intra-cranial pressure, the neuritis would already have subsided. But any good it can do must depend on what capacity for recovery remains in the fibres of the optic

---

* See *Postscript*.

nerve. In optic neuritis the nerve fibres may be damaged by propagation to them of inflammation, also by the interstitial contraction of the products of inflammation when these are abundant. At the beginning of subsidence, if there is but slight loss of sight, or even if there is none, vision may almost entirely fail during the subsidence, in consequence of the damage by the contraction of the inflammatory tissue between the fibres. If the neuritis has subsided almost to the level of the retina, and there is still some sight, it may be improved by the recovery of the functional power in fibres which are still continuous, although narrowed. But if sight is gone before subsidence begins, it is certain that the subsidence will complete the destruction of any fibres that remain, and no appreciable recovery of sight is possible in such cases.

In our patient there is still in the right eye a little perception of light, but there is much contraction of tissue still to go on, and I think that before that subsidence reaches the retina, sight will become extinct.

Postscript.—The lecture is left as it was delivered, although many statements have been contradicted by the later facts of the case. But it is often useful thus to perceive such traversing of inference and fact, because it discloses the limits of our power of discerning that which is, and of forecasting that which is to be.

After the lecture was delivered, the extensive swelling in the right eye lessened with a rapidity that I have never before observed; a difference could be perceived daily, and in a week the swelling was not wider than twice the diameter of the disc. The subsidence seems to show that the swelling was due to œdema in larger degree than I conceived. The part played by œdema has been illustrated in one of the microscopical figures in "Medical Ophthalmoscopy," and the case gives increased support to Mr. Marcus Gunn's inferences in the paper referred to. With this subsidence the headache became much less for several days. It then increased to more than its old intensity, although the subsidence of the neuritis continued. Trephining, postponed on account of the improvement, became compulsory. It will be remembered that the seat of the pain and its passage down the neck, with retraction of the head, pointed to subtentorial disease, but the tenderness to percussion was distinctly just *above* the tentorium on the left side. Mr. Horsley trephined below the tentorium at the left occipital bone, adopting the method he finds best, first to trephine, and in a subsequent operation to divide the dura mater. In this case he proposed, if he found no subtentorial disease, to divide the tentorium and explore above it. After the trephining, before the second operation that he intended, the patient died, with general convulsions, coma, and respiratory failure. After death an enormous cyst was found above the tentorium, in the left occipital lobe. It contained at least five ounces of clear, pale fluid, and had pushed forward the ependyma lining the posterior horn of the lateral ventricle, from which it was completely shut off. There were no crystals of any kind in the fluid.

To the eye the cyst seemed simple; no growth could be discerned. But such a simple cyst is almost unknown; for a simple cyst to cause the symptoms of aggressive irritation is, I think, quite unknown. Microscopical examination of the wall, by our able pathologist, Dr. Colman, showed at one place, at the posterior part of the floor of the cyst, distinct evidence of a morbid growth. There was a layer about 2 mm. thick of sarcomatous tissue, consisting of uniform round cells, with abundant embryonic blood-vessels. In several points the growth could be seen to be invading the white matter along the lines of the vessels. A considerable sized artery in the neighbourhood was completely thrombosed, and there was marked hyaline degeneration of the blood capillaries in the vicinity of the growth.

The chief practical lessons the case teaches are these: (1) That distinct local tenderness has relatively a very high localising value in cortical lesions. (2) That rapid subsidence of an unusually extreme swelling of the optic papilla may not have the significance that is possessed by slower subsidence of moderate papillitis. Any initial hemianopia was not observed, and when the patient came under observation sight had failed too much for it to be found.

Yet one other fact remains to be mentioned, of all, perhaps, the most anomalous. If ever the ophthalmoscopic changes suggested increased intra-cranial pressure, it was in this case. If ever a morbid process seemed capable of causing increase of intra-cranial pressure, it was in this case. Yet the condition within the eye, suggestive of the cause, subsided rapidly, while the cause remained, as far as could be afterwards discerned, unchanged. Lastly, the mechanism by which intra-cranial pressure seems to act is considered, on good ground, to be always revealed by the distension of the sheath of the optic nerve behind the eye—enlargement of the space continuous with that beneath the arachnoid. In this case there was no distension of the sheath, and there was not the least evidence of past distension.—W. R. G.

www.ingramcontent.com/pod-product-compliance
Lightning Source LLC
Chambersburg PA
CBHW031931230426
43672CB00010B/1882